# SCIENCE IN PUBLIC

# SCIENCE IN PUBLIC
## Communication, Culture, and Credibility

JANE GREGORY AND STEVE MILLER

BASIC
BOOKS

A Member of the Perseus Books Group

Library of Congress Catalog Card Number: 00-105208
ISBN-10: 0-7382-0357-2   ISBN-13: 978-0-7382-0357-7

Basic Books is a Member of the Perseus Books Group.
Find us on the World Wide Web at http:// www.basicbooks.com

Books published by Basic Books are available at special discounts for bulk purchases in the U.S. by corporations, institutions, and other organizations. For more information, please contact the Special Markets Department at the Perseus Books Group, 11 Cambridge Center, Cambridge, MA 02142, or call (800) 255–1514 or (617) 252–5298.

First paperback printing, July 2000

# Contents

# Contents

# Contents

# Preface

**Science has been very much in the public eye in recent years. Not only that,** but scientists themselves have had the public very much in mind. Phrases such as "public understanding of science" and "scientific literacy" have achieved unprecedented currency. Both public understanding of science and scientific literacy have been deemed to be "a good thing," and demands have been made—by scientists, politicians, and educators—for more of them.

What the public think of science and what scientists think of the public, and how the media bring the two together, have been matters of some research and considerably more opinion. Sweeping claims are made about the complex interactions of science in the public, and agendas—social, political, and academic— are rife. Despite the long history of these interactions, there are many people who convey the distinct impression that now is the first time Western culture has attempted to deal with science in public.

It may be fair to say that now is the first time in history when public understanding of science has been analyzed to any great extent. The "problem" of science in public is now an object of study for academics. Their efforts, drawing on long and respected methodological traditions and informed by painstaking research, have served to highlight the complexity of the entities involved; but they have also begun to unravel some of these complexities and to offer paths along which one might achieve some understanding of the public understanding of science.

A considerable part of our motivation in writing this book was our own frustration at the sloth with which the results of this work seem to be informing policies and practice in the public understanding of science, and at the frequency with which commentators in what is in essence an interdisciplinary field ignore (or sometimes, even actively condemn) the work of those in other disciplines.

If the field has any coherence at all, then it is to be found in its central aim: that of somehow improving the relationships between science and the public. "Improving" here means different things to different people, and herein the tension lies; but a rapprochement of some sort is what all parties who take an interest want. It seems ironic to us, then, that the same parties find it quite so difficult to come together themselves, learn from each other, and move forward. Many different groups have invented wheels now: it is time to build the bus.

What we offer is a synthesis. The ideas of sociologists and communications researchers rub shoulders here with the promises of politicians and the hopes of educators. The public are here, and so is science, in both their idealized and real-world guises. The book's scope is broad, but so is its subject; getting the perspectives of scientists, historians, psychologists, sociologists, communications researchers, lay groups, and journalists into a single book has been difficult. We hope we have done the people and their ideas justice, and we hope that in selecting the people and their ideas we have provided readers with some points of access into these fields. Most of all, we hope that the stories and theories of science in public will conspire here to inform and fuel the continuing debate about public understanding of science. Almost everyone will find something to disagree with; we look forward to what we hope will be a constructive debate.

In writing this book, we have been assisted enormously by discussions with friends and colleagues, at University College London, Imperial College, London, the London School of Economics, the London Science Museum, the Wellcome Centre, and Cornell University, as well as many elsewhere. We would particularly like to thank Martin Bauer, Brian Balmer, Robert Bloomfield, Joe Cain, Cathy Campos, Hasok Chang, Robin Clegg, Lisa DeProspo, John Durant, Steven Epstein, Colin Finnery, Andrew Gregory, Sally Gregory, Rebecca Hurst, Frank James, Helene Joffe, Bob Joseph, Simon Joss, Marina Joubert, Sandra Jovchelovitch, Bruce Lewenstein, Marion McDowell, Arthur Miller, Norma Morris, Akiro Nishimura, Asdis Ragnarsdottir, Michael Rodgers, Bonnie Schmidt, Jonathan Tennyson, Geoffrey Thomas, Jill Turner, Jon Turney, Barbara Ward, and Maurice Wilkins. One of us (S.M.) would like to thank Mary, Sean, Lisa and Simon for support and understanding during a year when he has not been the nicest person to know.

Jane Gregory
Steve Miller

## Chapter 1

# The Recent "Public Understanding of Science Movement"

**In the last decade or so, scientists have been delivered a new commandment** from on high: *thou shalt communicate.* In the recent past, many scientists looked at involvement in the popularization of science as something that might damage their career; now, they are being told by the great and the good of science that they have no less than a duty to communicate with the public about their work. There are even cash inducements, from agencies funding scientific research, for scientists to popularize science. Previous generations of young researchers had become used to being told that their place was in the laboratory: they were often brought up in a culture which said that science—if it were good science—should be generally unintelligible to all but an elite. Now, many of the new generation of student scientists are being coached in communication skills to equip them for talking intelligibly to the outside world.

For their part, the public—those outside of the scientific elite—had previously to marvel at such wonders of science as they were allowed to behold, to be grateful for the benefits scientific advances brought to society, and to be just a little frightened of scientific knowledge—at least frightened enough not to meddle or to voice their uninformed opinions on scientific matters. From time to time, to make sure things did not get out of hand, the fears of the public would be assuaged by the reassuring figure of the expert, who did know enough to probe the innermost secrets of animate and inanimate nature. But now the scientific establishment and national governments insist that the public must understand science if they are to be useful citizens, capable of functioning correctly as workers, consumers, and voters in a modern technological world. The media, under the direct control of neither scientists nor government, have nonetheless been put under pressure to communicate more science.

The Movement for Public Understanding of Science has arrived. It makes new demands on busy scientists, journalists, and the public. And, according to its most enthusiastic supporters, it opens up unparalleled possibilities in the realms of commerce, culture, and democracy.

But this movement is not without its critics. Fears are often voiced of an overweening scientism, mechanically reducing every area of human experience to those of its components about which science has something to say. There are concerns, too, that the whole public-understanding-of-science enterprise is motivated solely by scientists' ambitions to drain yet more cash from the already overburdened taxpayer. And there are fundamental objections that the activities that purport to increase public understanding fail badly both to address the real needs of citizens and to help them make sense of their world.

This book aims to explore the basis on which efforts to improve public understanding of science are founded and to look at some of the historical roots of those efforts. It examines the idea of public understanding of science and the notions of public, understanding, and science embedded within it. It assesses the objections and obstacles to attempts to improve the public understanding of science. Since the media play such an important role in the relationship between the scientific community and the public, this book looks at issues and examples of science in the mass media. It also analyzes the interactions in the public sphere when science and the public meet. The book concludes with a protocol drawn from these discussions, which, by relating theory to practice, aims to further constructive mutual relations between science and the public.

The work of many authors from many different fields is represented here. We hope that our survey is at least representative of the wide range of efforts that scholars from a variety of academic traditions have made in their attempts to decipher the complexities of the area. The potential scope for such an effort is vast, but we hope that this book will at least serve as a useful introduction to science in public and that it will point those readers who wish to go further in interesting and fruitful directions.


## SCIENCE AND THE PUBLIC

Caricatures of the relationship between science and the public are legion. Scientists are both mad and the only sane and rational people on the planet; they are loved by the public as saviors and hated as designers of weapons of mass destruction; they are derided as "nerds" and revered for their single-minded pursuit of truth. Separating the fact from the fiction is difficult. Caricatures draw their power from their proximity to the truth; however, as this book aims to show, the relationship between science and the public is much more complex and nuanced than any caricature can convey.

If the media are a more or less accurate reflection of their audience's attitudes, then, immediately after World War II, science and scientists were generally held in high regard by an interested public. Three decades later, however, with concerns about industrial pollution and civil and military nuclear issues high on the agenda, public attitudes to science were far more ambivalent.[1] At the same time, governments throughout the world were looking to trim budgets, including those for research. With science under threat, scientists around the world began to wonder if the relationships between science and the public perhaps needed a little attention; and their efforts, along with the reactions of social scientists and the activities of politicians, educators, and interest groups, have come together under the banner of public understanding of science with such coincidence and cooperation that both advocates and critics alike speak of a public understanding of science "movement," which is manifesting itself, in different ways and for different purposes, in many parts of the developed and developing worlds.

For many activists in this movement, a useful rapprochement could be achieved if only the public could think like scientists. The idea has some precedents: in the 1930s, the U.S. educator John Dewey claimed that young people should be inculcated with a "scientific attitude" that would help them approach the issues and problems of everyday life in a rational and logical fashion.[2] This struck a chord in the United States between the wars: the power and authority of science was co-opted by politicians across the political spectrum, and the ideas of science were a prominent part of intellectual and popular culture. Scientific institutions had wised up early to the power of the media and, in 1920, had launched a science news agency, the Science Service, to bring science and scientific ways of thinking into public view. During World War II, the military achievements of science gave it a moral stature that assured it a prominent role in peacetime, and one that scientists, politicians, and publishers were quick to exploit.

The American Association for the Advancement of Science (AAAS) was reminded of its long-standing constitutional commitment to the public understanding of science in 1951, when board member Warren Weaver called attention to "the broader external problem of the relation of science to society":

> . . . the AAAS [should] now begin to take seriously one statement of
> purpose that has long existed in its constitution . . . : to increase public
> understanding and appreciation of the importance and promise of the
> methods of science in human progress. . . . in our modern society it is
> absolutely essential that science—the results of science, the nature and
> importance of basic research, the methods of science, the spirit of sci-
> ence—be better understood by government officials, by businessmen,
> indeed by all the people.[3]

Weaver's statement prompted some realignment of AAAS policy, but it also created considerable unease within the Association: scientists were concerned that
greater engagement with the public might dilute the Association's commitment to
science. In 1955 Weaver was President of the AAAS, and in his address to its annual meeting, he stressed the benefits to science:

> It is hardly necessary to argue, these days, that science is essential to
> the public. It is becoming equally true, as the support of science moves
> more and more to state and national sources, that the public is essen-
> tial to science. The lack of general comprehension of science is thus
> dangerous both to science and the public, these being interlocked
> aspects of the common danger that scientists will not be given the free-
> dom, the understanding, and the support that are necessary for vigor-
> ous and imaginative development.[4]

In practice, the AAAS took some time to rally to Weaver's cause, just as science
educators were slow to implement Dewey's project. The change in attitudes came
in part in reaction to a huge blow to American pride: in 1957, at the height of the
Cold War, the Soviet Union launched the first satellite to orbit the Earth successfully. Sputnik fascinated the world and disconcerted the West, and Americans in
particular felt the time had come for an urgent implementation of Dewey's ideal.
A survey carried out by the National Association of Science Writers (NASW)
showed that while American attitudes toward science were generally positive, levels of factual knowledge of science were low.[5] Consequently, over the next two
decades, the U.S. federal government authorized several billion dollars to be spent
on science education in public schools: the rationale was that knowledge of science would have to be given by the formal education system if Americans were to
cope with the demands of, and compete internationally in, a modern technological
society.

Systematic attempts to determine whether these funds were having any effect on levels of what came to be known as "scientific literacy" began in 1972,
when the National Science Board commenced its biennial *Science Indicators*—
social surveys that gauge people's knowledge and understanding of, and attitudes

to, science. According to political scientist Jon D. Miller, who has administered and analyzed the surveys since 1979, the data that year, for example, indicated that only 14 percent of Americans could give a satisfactory answer when asked what it meant to study something scientifically—a satisfactory answer being one that included the notion of testing and modifying hypotheses by, and in the light of, experiment.[6] This number was not significantly different from the 12 percent who could achieve this level of scientific literacy in the NASW's 1957 survey. And when those of 1979's 14 percent who "failed" other tests were discounted (such as those who believed that astrology was at least "sort of" scientific), the overall percentage of scientifically literate Americans was halved to 7 percent. In 1985, the figure was 5 percent.

The low percentages of Americans achieving a "satisfactory" level in survey after survey have led science educator Morris Shamos to question the desirability of achieving universal scientific literacy among his compatriots[7] and have led others to suggest it to be downright impossible.[8] We examine their arguments in Chapter 4. Another point of view is that the test criteria are simply too stringent: the 1985 survey required that respondents show understanding of the "scientific process," of "basic terms and concepts," and the "impact of science and technology on society" in order to be classed as scientifically literate. The difficulties with this type of definition of scientific literacy are illustrated by the survey carried out in Britain in 1988 by John Durant, Geoffrey Evans, and Geoffrey Thomas.[9] When asked the abstract, open question "what does it mean to study something scientifically?", only 11 percent of respondents gave an answer that involved the idea of experimentation, while just 3 percent mentioned testing hypotheses. But given a choice of techniques for evaluating the effectiveness of a new drug—that is, when presented with essentially the same question expressed as a concrete example—more than half chose the answer that involved using experiments to test hypotheses.

The 1988 British survey was just one outcome of a surge of interest and activity in the public understanding of science in Britain that was prompted by a 1985 working-party report from the Royal Society—Britain's premier scientific institution. The report's terms of reference were "to review the nature and extent of public understanding of science in the United Kingdom and its adequacy for an advanced democracy; to review the mechanisms for effecting public understanding of science and technology and its role in society; [and] to consider the constraints upon the processes of communication and how they might be overcome."[10] The working party, which was chaired by geneticist Sir Walter Bodmer and included eminent scientists as well as the film maker David Attenborough and sociologist of science John Ziman, took evidence from a broad range of journalists, scientists, industrialists, and politicians—from all concerned parties except the general public. Among the conclusions drawn were that everyone should have some understanding of science, preferably provided initially by their school education, and that there would be occasions when this would help them make better personal decisions. The report singled out the importance of the Parliamentary and Scientific Committee (of Members of Parliament) in discussing the scientific aspects of issues to be debated in Parliament and suggested that those who led British industry needed a better understanding of science if the country was to remain economically competitive. The report emphasized the need for more science in the media, and, addressing its own community, it concluded:

> Scientists must learn to communicate to the public . . . and . . . consider
> it their duty to do so. The Royal Society should make improving public
> understanding of science one of its major activities.

One result of this report was the formation of COPUS—the Committee on the Public Understanding of Science. COPUS is a collaborative effort between three long-standing pillars of Britain's scientific establishment, the Royal Society, the Royal Institution, and British Association for the Advancement of Science, which all have strong historical pedigrees as promoters of science to the public. COPUS set in motion a number of projects, including an annual science book prize, small grants for public-understanding-of-science activities, and media training workshops for scientists. The report also prompted a research program, funded and coordinated by the Economic and Social Research Council, in which a number of research groups from different disciplines in the social sciences and humanities undertook quantitative and qualitative research on the public dimensions of science and technology. Much of the British research discussed in Chapter 4 has its origins in this program.

Public understanding of science has in recent years become part of official government thinking in the United Kingdom. In 1993, a White Paper entitled *Realising Our Potential* stated: "The aim is to achieve a cultural change: better communication, interaction and mutual understanding between the scientific community, industry and Government Departments."[11] One means of achieving this would be by working with funding agencies to "improve scientists' understanding of communicating with the public," and there would consequently be an "emphasis on communication skills in research training," especially for graduate students. Since 1994, a government-sponsored "Science, Engineering and Technology" activity week has brought science to the attention of many people who might otherwise have ignored it, not least government ministers themselves who have been publicly involved with several events. Many of the sentiments of *Realising Our Potential* were echoed and amplified two years later in the report of a Cabinet Office committee under the chairmanship of former Astronomer Royal Sir Arnold Wolfendale.[12]

While the main thrust of the public-understanding-of-science movement in Britain has been toward adults, families, and community groups, either directly or through the media, its counterpart in the United States has, as implied by the term "scientific literacy," retained its educational slant and focused on school and college students. In Britain too, there has been national curriculum reform that prescribes that all school students should study science at least until the age of 16; but in the United States it is the major scientific institutions that are providing the educational impetus. The AAAS's "Project 2061" aims to "reform . . . education nationwide so that all high-school graduates are science literate." Panels of scientists decided on the knowledge and skills students should acquire during their 13 years at school, and Project 2061 (the date is of the next visit of Halley's comet) offers educators resource materials to help them build the achievement of scientific literacy into the curriculum. The National Association of Science Teachers' "Scope, Sequence and Co-ordination of High-School Science" scheme, instituted with the support of the National Science Foundation, aims to facilitate science education and encourage individual school districts with differing policies on science to meet national standards. The National Science Foundation's activities in sponsoring science communication in the mass media and in community projects come under the budgetary heading of "informal science education," while the U.S. federal government has provided funds for educational campaigns in areas such as diet and food safety. The AAAS has given awards for science journalism and for popularization of science by scientists, produces radio programs on science, and sponsors science activities for minority groups. Since 1959, it has hosted COPUS&T, a committee that organizes meetings, promotes activities, and facilitates discussion on the public understanding of science.

The decade since the establishment of COPUS in Britain and Project 2061 in the United States has seen many new ventures in the broad field of the public understanding of science in many different forums around the world—indeed, they amount almost to a public-understanding-of-science industry, which is colonizing small corners of academia, commerce, and politics and generating its own momentum. Its products include clearer perspectives on previously ill-defined or even undetected problems; policies and activities designed to deal with these problems; and manifestos and protocols designed to redesign professional and educational practices. And—a sure sign that the area has academically come of age—there are two international peer-reviewed journals dedicated to research into all aspects of the public understanding of science.[13] A key word in this industry is "initiative": in research, practical efforts, and mission statements, the movement for public understanding of science is rife with initiatives. There is a strong sense among the most vocal of the public-understanding-of-science industrialists that, in the closing years of the 20th century, science and the public are starting afresh.

## WHAT DO WE MEAN BY SCIENCE, UNDERSTANDING, AND THE PUBLIC?

Not everyone is quite so gung ho about the public understanding of science. There are scholars who are not convinced that the public understanding of science needs improving and who are not quite sure who, if anyone, would benefit if it were improved. There is considerable disagreement about what "improvement" might mean in this context. At an even more basic level, there is much discussion about what the terms "public," "understanding," and "science" actually mean. All in all, disagreements abound as to whether the public (whoever they are) should, can, or would want to understand (whatever that means) science (however defined).

The public-understanding-of-science movement has not overly concerned itself with articulating its understanding of science: science is taken as given. Other scholars, however, from the fields of history, philosophy, and sociology of science have made important contributions to recent efforts to understand the public understanding of science by examining science itself. As relative outsiders to science, these researchers have drawn attention to many aspects of science that scientists themselves usually take for granted, and in so doing they have revealed some of the particular challenges that the knowledge, practices, and organization of science present to the public understanding of science. We look at these scholars' sometimes controversial work, and at scientists' reactions to it, in Chapter 3.

The public has not been examined too closely either, but for a different reason: the public turns out to be rather too complicated. A quick fix is to think of the public as a black box—and as an empty black box at that. Again, though, other scholars have contributed to the debate: the public has been the subject of sociological inquiry for some time, and we look at some of the results of this work in Chapter 4.

The public understanding of science movement has, however, had much to say about "understanding." Some commentators equate "understanding" with "knowledge." Following social commentator Edward Hirsch's 1987 attempt to set out the requirements for Americans to be culturally literate,[14] others have entered the fray to focus on science. Science writer Richard P. Brennan produced a *Dictionary of Scientific Literacy*, aimed at the intelligent nonscientist, which includes over 600 items ranging from "absolute zero" to "zygote."[15] Geophysicist Robert M. Hazen and physicist James Trefil warned that Americans faced with newspaper headlines that directly affected their lives—such as "Genetically Engineered Tomatoes on Shelves" or "Japanese Take Lead in Superconductor Race"—would need to "fill in whatever blanks have been left by [their] formal education" in order to understand the issues involved.[16] Among the fillings they offered for these blanks were "covalent and ionic bonds," "the gauge particle," "jovian planets," "recessive genes," and "tRNA." Trefil subsequently outdid Brennan with *1001 Things Everyone Should Know about Science*, which included the information, for example, that all proteins found on Earth could be constructed from just 20 amino acids.[17]

Other commentators, however, equate "understanding" with "appreciation." In the wake of the 1993 U.S. Congress decision to shut down the Superconducting Super Collider project to build the next-generation particle accelerator—a decision many scientists blamed on public indifference to science in general rather than on a congressional desire to seek a more balanced federal budget—particle physicist Leon Lederman used his keynote address to a Chicago conference on scientific literacy to outline his remedy in terms of what the public should know. Lederman's list of nine distinguishing features of science that he thought should be generally understood included the nature of scientific concepts, models, and theories and how they can be tested and verified. Interestingly, it also included the idea that science only gives provisional answers to the questions to which it is responding and, not surprisingly in the circumstances, the suggestion that the public should realize just how costly it is to reach even the relative truths of science.[18]

There is no doubt that many in the scientific community who want to further the public understanding of science are really concerned with increasing the public's appreciation of science, with a view to enhancing the status of themselves and their colleagues. Research funding organizations that rely on public money have clearly aimed at ensuring that their budgets are protected and, if possible, augmented. But such scientists and their organizations have also often been in the forefront of those pointing out that scientists cannot simply make themselves available to the media when they want to glory in a new discovery: they must be on hand to take part in debates where they have useful expertise and to answer awkward questions of public concern. We look at ideas about the public understanding of science, and about the role of scientists in public issues, in Chapter 4.


# WHY THE PUBLIC UNDERSTANDING
# OF SCIENCE MATTERS

Despite the difficulties most commentators and scholars face when they attempt to define "public understanding of science," there are many people who believe that more of it—whatever it is—would be "a good thing." Science writer Isaac Asimov considered it to be essential to preventing growing public hostility toward, and suspicion of, scientists and their works. Enabling citizens to live in a modern world was one of the Royal Society's aims. And democratic governments are concerned with explaining to taxpayers just how their money is being spent, as well as wanting the economic advantages of a scientifically aware society. As Wolfendale's report summarized them: "The objectives of the [U.K.] Government's policy on public understanding are: 'to contribute to the economic wealth and quality of life of the Nation, particularly by drawing more of our best young people into careers in science, engineering and technology'; and 'to strengthen the effectiveness of the democratic process through better informed public debate on issues of public concern arising in the fields of science, engineering and technology'."[12] U.S. Vice President Al Gore told the AAAS in 1996 that "we need to create a learning society . . . that harnesses the power of distributed intelligence and uses it to lift our lives."[19]

One of the most succinct summaries of reasons for enhancing public understanding of science was produced in 1987 by Thomas and Durant.[20] They grouped the various motivations under nine "benefits" that are supposed to accrue to society, organizations, or individuals. It is worth following their arguments in detail, since they provided a framework for many further studies. Their scheme runs as follows.


## Benefits to Science

While most scientists do not anticipate the hostility and suspicion that Asimov predicted, many believe that science—its institutions and individual scientists—has much to gain from a wider public understanding of what it can contribute to practical and cultural life. Science needs at least tolerance from the broader community if it is to function, as those branches of science that some sections of the public will not tolerate know to their cost.

Much scientific research is funded out of taxes, and those who pay also vote and put pressure on their representatives. Indeed, the cynical view is that the whole public-understanding-of-science enterprise is nothing more than an exercise to ensure continuing government funding of research: the more people know about science (by which scientists tend to mean "the more people appreciate science"), the more money will be given to it. It seems not to occur to the proponents of this argument that, were the chance to arise to vote on single scientific issues in our imperfect democracies, the public has an equal opportunity to vote against science.

It may go against the grain of centuries of tradition, but scientists and their official bodies can claim just as much right to lobby for their interests as any other professional or special-interest group. Certainly, the American space agency NASA mounted a huge public relations exercise in August 1996, when its scientists published results that seemed to show that trace chemicals and structures in a Martian meteorite were fossil remains of life. The announcement—even with all the caveats surrounding it—was timely, since there were several planned missions to Mars for which federal funding was very far from secure. NASA was rewarded with an endorsement by President Bill Clinton. Promising to follow up on the discovery by funding further missions to Mars, Clinton said that if the detection were confirmed, "it will surely be one of the most stunning insights into our world that science has ever uncovered." Clinton's 1996 election campaign promised to use science to build "a bridge to the twenty-first century."[21]

But it does not always work like that. As we have seen, the U.S. Congress closed down work on the Superconducting Super Collider, despite having already sunk $2 billion into its construction. In the run-up to that decision, however, Steven Weinberg and Leon Lederman—two of the country's Nobel laureates in physics—had written books aimed at explaining to the public just how important the project was.[22] Either not enough people read the books to produce a sufficient political head of steam to save the project or those who read them were not convinced by Weinberg's and Lederman's arguments.

Which brings us to another question about how much science really benefits from greater public awareness. Underlying much of the reasoning of the public understanding of science movement is the notion that greater knowledge leads inevitably to greater appreciation. But the 1996 British Social Attitudes survey showed that while knowledge of science had increased significantly since the 1988 Durant, Evans, and Thomas survey, overall attitudes toward science had hardly changed at all; if anything the public were more ambivalent. There is some evidence that individuals who are more knowledgeable about science do have more positive attitudes toward it.[23] But at a certain level, a greater understanding of science and its social implications can lead to criticism and even hostility. One has only to think of nuclear energy and its associations with planned or accidental mass destruction and chronic radiation effects[24] or the new advances in genetics and fertility treatment that are raising again the specter of Frankenstein.[25] Even where—as often in the case of environmentalists—critics make use of scientific knowledge and argument in pursuit of their case, they can be in sharp conflict with the scientific establishment.

## Benefits to National Economies

While many have argued that science directly benefited industry very little prior to World War II, there do appear to have been dramatic changes since then. The nuclear and modern biological industries are clear cases in point, and scientific developments such as lasers and transistors have clearly revolutionized modern technology. From this it follows that a steady stream of high-level researchers and technically accomplished industrial workers is required for any country wishing to compete internationally.

However, governments in both the United States and Britain do not seem to have known quite how to capitalize on the world-leading research done by their countries' scientists, particularly where turning fundamental research into salable products is concerned. The example of Japan is often quoted—supposedly a country that has come to the forefront of the world economy on the basis of the most up-to-date technology without carrying any of the burden of the original fundamental research. In Britain, in particular, this has led to significant swings in policy: first, injunctions from government that research councils should be funding "near market" research, ready to hand over to industry; then an insistence that fundamental "blue skies" research is the thing—coupled with blandishments to industry to put more of its cash into product development.

Other economic arguments for wider understanding of science concentrate on the public as consumers. If a strong economy depends on a strong home market, then consumers need to be scientifically aware in order to appreciate what is being offered for sale. Of course, the benefit accrued to the economy of any individual nation depends on the extent to which the home economy can supply the needs of these new techno-consumers.

## Benefits to National Power and Influence

In a world where countries are interdependent both economically and politically, but at the same time divided into various more or less antagonistic blocks, the ability to wield power and influence on the global stage is extremely important. In particular, where two or more opposing superpowers are trying to convince their neighbors that theirs is the best way to organize society, proven technological and scientific superiority can be very persuasive. But ensuring that there are resources to achieve this superiority means convincing at least a part of the nonscientific community in any society that the requisite technology is worth the investment. Hence there is a link between the public appreciation of science and the ability of a nation to be internationally influential.

Arguably the most visible icon of national scientific prestige in the decades since World War II has been the technology of space travel and exploration. No one was more disturbed than the Americans by Sputnik; according to Alan Shepard and Deke Slayton, founder members of NASA's first space team: "American skies had been violated. During World War II no Nazi or Japanese aircraft had penetrated US air space. But here was Sputnik, made in Russia, passing overhead several times a day. The Eisenhower administration was caught by surprise."[26] From then on, the U.S. government spent billions of dollars on the "space race." NASA eventually had the satisfaction of landing the first man on the Moon, "demonstrating" the superiority of the United States over its rivals.

High-profile successes can too easily give way to high-profile disasters, however. Science can contribute to national shame as well as prestige. Commenting on the 1987 *Challenger* shuttle disaster, science writer James Gleick portrayed the event as a symbol of "machinery out of control":

> The American space agency had made itself seem a symbol of technical prowess, placing men on the Moon and then fostering the illusion that space travel was routine—an illusion built into the very name shuttle . . . the space shuttle explosion seemed a final confirmation that technology had broken free of human reins. Did nothing work any more?[27]

## Benefits to Individuals

The COPUS report pointed out that "many personal decisions . . . would be helped by some understanding of the underlying science." This theme was taken up by Thomas and Durant, who suggested that "more knowledgable citizens are able to negotiate their way more effectively through the social world."[20] So the idea that understanding more about science can benefit people who are making sense of their everyday lives is one claim advanced by the public understanding of science movement that is not just about institutions and interest groups.

There is evidence that when people are directly affected by scientific matters—for example, in the case of chronic or inherited disease—they become very knowledgable about the complaint that affects them and their families.[28] Workers in some science-based industries, too, are strongly motivated to understand what they are dealing with. But the evidence is much less clear whether, for example, women genetically prone to breast cancer understand their situation in terms of the science of genetics rather than family experience.[29] And workers in the nuclear industry understand radiation as a health and safety issue rather than as alpha, beta, and gamma rays; they often choose to trust their colleagues to have provided a safe environment for them, rather than engaging with the technical issues themselves.[30] For the rest, most people make use of electrical equipment on a pragmatic basis, without resort to Maxwell's equations of electrodynamics or semiconductor theory.

The other main claim for an understanding of science among the public as individuals concerns the job market. A large proportion of the armies of unemployed in Western economies are unskilled young people, unqualified for what job opportunities do exist. This has led governments to make a series of changes to the education system in attempts to ensure that every school student studies enough science to fit them for the increasingly technological job market.

## Benefits to Democratic Government and to Society as a Whole

Many of the issues that come up for political discussion involve science. Several authors have described an "information society," where possession of information is power.[31] Informed voters are supposed to be able to exert pressure through the ballot box and by lobbying. Informed consumers exert their pressure by buying or boycotting. But with elections spaced on timescales of years and with pricing and display policy almost totally under the control of producers and retailers, the weapons in the hands of ordinary citizens are fairly blunt. Little has been done that would really turn greater public understanding of science directly into democratic political and economic power. Thomas and Durant also caution that, even if this could be achieved, "wider understanding . . . will not automatically lead to a consensus."[20]

One major social concern is the creation of a group that has no work, no prospects, and thus no stake in the present or future of society. Such bleak prognoses have their echo in the area of public science, where an increasing proportion of the population—of all social classes and intellectual abilities—is being left behind by the onward march of scientific discovery. This group is, according to some analyses, becoming a scientific underclass, increasingly unable to understand the modern world, and thus unable to function as citizens of a technological society.

The historian of science and physicist Gerald Holton also points out that one of the key groups in what he terms the "anti-science movement" is influential but alienated intellectuals who would normally have felt themselves allies of science and progress. Instead, they find they share a feeling of "humiliation" at their inability to understand the latest developments of science. Shut out from the discussion of what appear to be important issues, they represent a dangerous intellectual polarization that threatens the integrity of modern thought. It is to counter such a dangerous split in society that public understanding of science should be improved.[32] We look at Holton's arguments in Chapter 3.

## Intellectual, Aesthetic, and Moral Benefits

There is no doubt that science and its attendants have inspired many cultural achievements, from the introduction of perspective in painting to literary triumphs such as Mary Shelley's *Frankenstein* and the whole genre of science fiction. However, at least in the Anglo-Saxon tradition, the place of science in culture, in intellectual life, is still problematic. In the late 1950s, novelist C. P. Snow pointed this out in his essay *Two Cultures*, which drew attention to a growing gulf between literary intellectuals and scientists.[33] While much criticized, Snow's broad-brush picture remains a topic of debate nearly four decades later, and is discussed in more detail in Chapter 2.

The Victorian poet Matthew Arnold valued science insofar as it was written and therefore part of literature, capable of being included in his definition of culture—"the best which has been thought and uttered in the world."[34] But he had no place in his culture for the scientific achievements of science *qua* science. And scientific training as a form of education would produce only "a useful specialist," not a truly educated man. Against this cultural snobbery, scientists have long argued not only that science is one of the greatest cultural achievements of the Western world, but also that its intellectual training is superior to that of the traditional arts and classics, as it produces a rigor they lack. Greater general public understanding of science, and particularly of its methods, would serve as an intellectual stimulus to the whole of society.

But would more scientific awareness and understanding lead to a less philistine society, one in which aesthetics and morals were held in higher regard? Isaac Newton's "collecting pebbles" metaphor for scientific discovery retains all the charm and wonder of the small child's search for the prettiest stone. More recently, images from the Hubble Space Telescope have found their way into art galleries as highly collectable objects. It has also been argued that scientists' quest for objective truth, recognizing the achievements of even their bitterest rivals in a fair and open-minded fashion, could serve as a moral template for working out the most intractable social and political disputes.

Thomas and Durant were quite well aware, however, that every positive reason they outlined was open to criticism and counterexamples; indeed, many of the criticisms predate their work. Among them are the opinions of those such as the novelist D. H. Lawrence, who saw in the explanatory power of science a grinding destruction of our ability to appreciate the art in nature—the science of optics has, in this thesis, stripped away the mysterious beauty of the rainbow. Worse still, the very open-mindedness of scientists—their readiness to accept that their knowledge is provisional, always open to modification by further experimentation and discovery—is responsible for introducing a moral relativism that is undermining the belief systems on which Western society is founded. In this view, science, far from being one of the most ethical pursuits open to people, is "spiritually corrosive."[35]

In 1952, the historian of science I. Bernard Cohen listed what he called the fallacies in arguments for improved public education in science.[36] These include the fallacy of scientific idolatry ("believing scientists to be lay saints, priests of truth, and superior beings who devote their lives to the selfless pursuit of higher things"); the fallacy of critical thinking (an understanding of science does not necessarily provide this faculty, as "may easily be demonstrated by examining carefully the lives of scientists outside the laboratory"); the fallacy of scientism (science is not necessarily the only or even the best way to solve many sorts of problems); and the fallacy of miscellaneous information ("the belief in the usefulness of unrelated information such as the boiling point of water . . . the distance in light years from the earth to various stars . . . the names of minerals"). These fallacies, and particularly the last, have cut no ice with those who administer knowledge surveys or write lists of facts—miscellaneous information—and publish them as the key to scientific literacy. Cohen called for the jettisoning of most of the facts of science—one can, after all, look those up in a book as and when needed—and stressed instead science for citizenship and "a feel for science." These are parameters that are difficult to measure using surveys, the traditional tool for assessing public understanding of science that the scientific community has been wielding since the late 1950s. Despite his criticisms, Cohen's agenda was pro-science: he looked forward to a day when journalists and educators would take up his challenge, for "that science can make the decisive contribution to the physical existence of man is without question."

Another critic is the sociologist Leon E. Trachtman, who in 1981 pointed out that while knowledge in itself might well be "a good thing," it is not always applied in everyday behaviors.[37] He offered as an example consumer behavior: consumers make decisions for many reasons, few (if any) of which may be scientific, and anyway it is difficult to find unambiguous scientific knowledge about how to behave among the claims and counter-claims of the marketplace. With regard to political behavior, Trachtman argues that

> . . . people who are consistently willing to make economic sacrifices in the interest of environmental preservation—or their opposite numbers—are unlikely to have their convictions and political activities modified by learning more about the life cycle of the Chesapeake Bay oyster.

Trachtman also charges that much science popularization, which, he suggests, represents science as "heroic, apolitical, and inherently rational," can produce "public misunderstanding of the nature, structure and character of science," which could in turn result in public disillusionment with science.

The public understanding of science movement of the 1980s has also attracted some new critics, particularly of the "deficit model" of public understanding of science, which sees the public as blank slates or empty vessels—as minds in deficit that need scientific information in order to be replete. The work of these scholars is discussed in Chapter 4: here let it suffice to say that their criticisms, along with those of Cohen and Trachtman which they echo, have barely impacted on scientists' discussions of public understanding of science.

# IS THERE ANYTHING NEW UNDER THE SUN?

We shall be returning in more detail to many of the issues and themes raised in this opening discussion. One question, in particular, leads naturally into the next part of this book—does the public understanding of science movement of the late 20th century in any way merit the label "initiative," or has it all been said before? In one sense, recent activity is an echoing of previous concerns, as we will see in Chapter 2; but it does seem to be special in its sheer scale and concertedness. There appears to be a genuine recognition on many sides that we cannot enter the next millennium with a society in which the "scientifically literate" are increasingly distanced from those who know nothing and care less about one of the great cultural achievements of modern civilization. For their own motives, government, the scientific establishment, and individuals have involved themselves in the public understanding of science movement in such a way that there remains room for dissident voices not only to be heard but also to influence the way in which the debate about science and its role in society will go. There is even a glimmer of an acknowledgment that all parties involved have a lot to learn about each other. As yet, we have no idea whether any of the current spate of activity makes any real difference to anyone. The recent episode in the public understanding of science may, like others, end up as mere fodder for the historians of scientific culture (and if so, this book may make their task a little easier); but, for the time being at least, the public-understanding-of-science industry believes that what it is doing really matters and is pushing its product to a public that, sometimes at least, seems to agree.

# Chapter 2

# Science in Public Culture

**Popular science is at least as old as science. Indeed, given that scientific activity** predates the establishment of a scholarly community for science, it could be argued that popularization pre-dates scholarly communication. It certainly long predates any "movement" for the public understanding of science. The first great writer of Western history, Herodotus (c. 484–425 B.C.), passed on to his public the Nile priests' explanations of how Egypt had been produced from silt deposited by the river—perhaps the oldest surviving popularization of geology. Four centuries later, the Roman poet Lucretius popularized the theories of the Greek atomists in his *De rerum natura*. Copernicus's *De revolutionibus orbium coelestium*, published in 1543 and one of the key texts of the scientific revolution, was arguably written for the general public. Certainly, Galileo's "crime," according to the Catholic church, was not that he had espoused the heliocentric model of the solar system but that he had popularized it. Even newspaper science has a long history. The first "science story" in an American periodical was published in 1690: it was a medical story about "plagues and agues."[1]

Despite this long tradition, popular science has only recently come to the attention of historians and cultural scholars. Rather like the music students who struggle to assemble libraries of the traditional songs that, as "low culture," were ignored by academics for so long, students of popular science are now attempting to recapture the ephemera of public scientific culture and to understand its role in the history of science and society. Some periods, and some players, have received more attention than others. This chapter makes no attempt at the difficult task of gauging the effect of popular science—if any—on the public understanding of science. But it does attempt a broad if episodic survey of the history of science in public culture in the United States and Britain, highlighting for closer inspection some events and practices that illustrate the spirit of their time and some scientist-popularizers who, for their techniques or for their motives, offer some insight into the culture of science in public.

## THE COMING OF AGE OF POPULAR SCIENCE

The social distincion between science and the public began with the formation of a community for science, that is, with the institutionalization of science as an activity with designated participants and with agreed rules and practices that separated it from other activities. A convenient date for the separation of the scientific community from the public at large is the 17th century: with the scientific revolution, science developed as an activity in its own right, and the Baconian enterprise needed champions. In England, the Royal Society was established in 1660. From the outset, it was a lobby group for science as well as an arena for scientific debate, and to that end, for the first one-and-a-half centuries of its existence, there were at least as many nonscientist gentlemen as active researchers among its membership. (Not until the 1820s did the Society start to take on its modern form as a mainly scientific forum, ejecting some of the gentlemen in the process.) From the outset, too, the Royal Society had its critics, fearful that this privileged body would, in its turn, privilege science above other cultural activities.

During the 17th century, scholarly works in natural philosophy in English began slowly to replace the more usual publications in Greek or Latin, and by the 18th century popular science was not only an intellectual but also a high-class social pursuit.[2] Later, improved printing techniques made books more available and more accessible to a wider group of people; and public lectures on science became a feature of cultural life, at least in the cities. Lecture-demonstrations encouraged participation, and scientific apparatus was manufactured and sold for home entertainment.[3] With the boom in international trading and merchant shipping, collections of exotica were exhibited in sophisticated homes and in public gardens.

The notion of "popular," however, presumes a "populace" or "public." The existence of this massive but previously invisible entity was made glaringly apparent to the British social elite by the French Revolution. The popularization of intellectual culture in France during the Enlightenment was held responsible for empowering the masses, and across the Channel social leaders took fright at the possibility of a repeat performance closer to home. The late 18th and early 19th centuries also saw important changes within science itself, with the establishment of the first dedicated scientific laboratories in which scientists "worked" rather than pursued their interests. In England, universities were slow to see the importance of science, but the Royal Institution was founded as a research laboratory in London, in 1799. One of its primary functions was to carry out studies for aristocratic "improving landlords," who wanted to apply the latest scientific knowledge to modernizing agricultural techniques.

The new Institution also had a key role to play in the popularization of science. Just as scientists were emerging from their private studies and into "public" laboratories (by now, scientists were not necessarily rich men funding their own work), so science popularization came out from the private salons and drawing rooms and into the lecture theaters that accompanied those public laboratories. In its first 50 years, the Royal Institution transformed the presentation of science to the public from the cosy after-dinner exposition and discussion to the professional scientific lecture, with the scientist–lecturer separated from his audience not only by the increased distance but also by the laboratory bench groaning under the weight of specialist apparatus.

The public lectures initiated at the Royal Institution by Sir Humphry Davy and developed by Michael Faraday (we look at Faraday's lectures in more detail in Chapter 6) proved extremely popular: the middle and upper classes found excitement and entertainment in the latest science; and skilled workers entered the lecture theater via a discrete, outside staircase, found a seat under the eaves, segregated from the gentlefolk, and drank in the knowledge that might improve their prospects in the new industries. This extension of popular scientific knowledge to the "lower orders" was not necessarily discouraged by their social superiors. The reforming Utilitarians, followers of the English philosopher and social commentator Jeremy Bentham, felt that an understanding of scientific matters would reinforce in the minds of workers the naturalness of the existing order and enhance social cohesion. By the 1820s, these thinkers were influential both in the Royal Institution and in the newly founded University of London (now University College London).[4] For their part, many skilled workers began to see the acquisition of scientific knowledge as a way of improving their lot not only as individuals but also as a social class—not for nothing did Marx later term his theory "scientific socialism." Public access to knowledge as a whole, and to science in particular, became a contentious political issue.

The politicization of science popularization in the 19th century coincided with new media and new publics: a growing urban class provided a market for cheap magazines and public lectures. Historian Robert V. Bruce reports that the booksellers of Milwaukee (population 3000) in the 1840s offered their customers popular works in botany, entomology, meteorology, and natural history, books by Dominique François Arago and John Herschel, and the complete works of Isaac Newton.[5] Increasing industrialization during the 19th century created another new audience: where once the landowners and entrepreneurs looked to traveling lecturers for new knowledge about steam engines and airpumps as an investment in a new industrial future, now the working men attended lectures, often given by distinguished scientists in the new Mechanics Institutes, that would make them qualified for technical jobs in the factories and mines. Magazines such as *The Athenaeum*, launched in London in 1828, reflected these trends.[6] Within a few years, its rather quaint, backward-looking attempts at popular science were replaced by full and businesslike reporting of the burgeoning scientific societies that accompanied the trend to "professionalism" and the separation of recognizably modern disciplines out of the rather undifferentiated natural philosophy of the 18th century.[7]

Science became a common amateur pursuit in the 19th century, on both sides of the Atlantic. Natural history was a popular pastime for men and women from many different walks of life. By the 1840s in the United States, even small towns might have a "scientific society" just as they might have a sewing circle or reading group. A rash of spectacular astronomical events in the 1830s and 1840s—comets, meteors, and the arrival of Neptune—sent the American public, with Herschel's best-selling *Treatise on Astronomy* in hand, out under the night sky. In Cincinnati, the groundswell of popular enthusiasm for astronomy encouraged some of the richer amateurs to fund a public observatory. The people's mandate did not, however, earn them telescope time for very long: after nine years of jostling for space, the professional astronomers closed the doors to the public in 1854.

Public lectures at the new Lowell Institute in Boston attracted huge crowds: in 1839, the stampede for free tickets for lectures on geology by Yale's Benjamin Silliman Sr. shattered the ticket-office windows. Because of the huge fees offered to speakers, thanks to an endowment by textile manufacturer John Lowell, the lectures also attracted distinguished speakers from far and wide, and some of them stayed on to make Boston a key center of U.S. science. One such was Louis Agassiz, who, as we describe in Chapter 8, was popularizing science in Boston long after his sensational public appearance at the Lowell Institute.[8]

In 1829, the Cambridge mathematics professor Charles Babbage started work on his *Reflections on the Decline of Science in England, and Some of Its Causes*, a theme that found an echo among many of his fellow scientists. His concern was that British science was falling behind its counterparts in other countries for lack of public interest. In part as a response to Babbage's call, the British Association for the Advancement of Science (BAAS) was set up, with the aim that is explicit in its title. The BAAS was intimately bound up with the politics of the popularization of science. Historians of science Jack Morrell and Arnold Thackray see its guiding principles as Liberal Anglicanism, in the sphere of religion, and centrist reformism, in politics: "The particular genius of the British Association for the Advancement of Science lay in its ability to serve as an instrument of public order and social cohesion," they comment.[9] After 1831, the BAAS staged an annual scientific jamboree attracting enthusiastic crowds of scientists and the public to hear lectures by the "great men of science" (until the turn of the century, they were exclusively men)—lectures that were reported at great length in the press. In the heyday of its scientific reporting, *The Athenaeum*, for example, devoted half of several issues to the BAAS annual meetings. It was at one of the early meetings that the polymath William Whewell coined the term "scientist" to describe the members of what was becoming an increasingly professional group. From the 1860s, the Association organized lectures for working men around Britain, with some spectacular successes: physicist Silvanus Thompson went to Cardiff, and miners organized excursion trains to his lecture; in Yorkshire, 3500 millworkers turned up to hear him.[10] In the United States too, leading scientists sought to inculcate the scientific worldview into the broader public and to equip the workforce for the new industrial economy, and touring lecturers preached self-help through scientific knowledge.[11] The American Association for the Advancement of Science was formed in 1847, with the central aim of drawing a clear line between professional and amateur science. At the AAAS meetings before the Civil War, scientists talked to each other in closed sessions during the day, and the public were admitted only to free lectures by big names in the evenings.[12]

Over the centuries, popularizers of science have had various intentions: in the first half of the 19th century, they wanted to bring to the masses the joy and moral benefit of knowledge; they wanted to reveal the hand of God in Nature; they wanted, by exposing the world as an organized, ordered system, to keep the working classes in their place. Under the guidance of men like Whewell and the reverend geologist Adam Sedgwick, English science, then weak in comparison with the standing it would acquire by the end of the century, sheltered under the protective wing of natural theology. This placed science in the same cultural environment as the arts and classics and made it suitable for inclusion in literary publications. *The Bridgewater Treatises*, which were published in the 1830s, gave eminent scientists of the day the chance to explore the connection between science and natural theology in greater depth. These books became bestsellers on both sides of the Atlantic. As the century progressed and science grew more assertive, however, the "common context" with natural theology—and thus with the rest of culture—began to fragment.[13] Physicist John Tyndall's address to the BAAS meeting in Belfast in 1874 exemplified the turn in this relationship. In his survey of the history of science, Tyndall talked of an ancient world in which the gods tossed all into confusion and in which science was "reined in" by Christianity. Science would slip these reins: "there grew, with the growth of scientific notions, a desire and a determination to sweep from the field of theory this mob of gods and demons, and to place natural phenomena on a basis more congruent with themselves."[14] Thus, in the latter half of the century, the purpose of popularization was much more clearly to bring to the public the intrinsic wonders of the latest scientific discoveries, rather than to demonstrate some higher purpose.

As well as controversies surrounding the relation of science to other areas of cultural and intellectual life, the public were treated to some of the most heated disputes between scientists. In the 17th century, such exchanges were often limited to letters, such as those between Newton and Hooke on the nature of light, which were then more widely read. The growth of popular outlets in the 19th century made these discussions more readily available to a much wider audience. The clash in 1860 between Darwin's bulldog Thomas Henry Huxley and Bishop "Soapy Sam" Wilberforce at that year's BAAS was a very public affair, and a source of joy to the cartoonists of the day. Lord Kelvin (while he was still Professor William Thomson) launched the first major blow in his war with the geologists over the age of the Earth in *Macmillan's Magazine* in 1862.[15] Kelvin's zenith in this affair occurred on the platform of the BAAS in 1894. His position was endorsed by BAAS President Lord Salisbury, who—much to the discomfort of Huxley, who was sitting alongside Kelvin—delivered a broadside at the theory of evolution, using Kelvin's estimate of 20 million years for the age of the Earth. Kelvin's nadir came too late for Huxley: in 1904, after the discovery of radioactivity and its implications for this debate, Ernest Rutherford strained diplomacy to the limit in a Royal Institution lecture that just about managed to point out that Kelvin's age of the Earth was wrong without quite damning the "Old Boy" as a fool.

Nor is mass media coverage of scientific controversies a recent phenomenon: in the 1890s, for example, publications such as the *New York Times* and *Harper's Weekly* filled many column inches with the technological competition occupying Tesla, Edison, and Marconi; Lord Salisbury's 1894 speech to the BAAS and the replies by Kelvin and Huxley were reported fully in the next day's London *Times*. While there was always room for celebration—Röntgen's discovery of X rays had made banner headlines in 1896—the coverage was not always uncritical: in 1899, for example, the U.S. magazine *Public Opinion* criticized scientists for making extravagant public claims for their own inventions. Certainly, not everyone in the 19th-century mass media felt obliged to take science completely seriously: *The Athenaeum* was quite happy to tweak the nose of eminent scientists; and in 1835 Richard Adams Locke wrote a series of front-page articles for the *New York Sun* about how John Herschel had discovered life—unicorns and winged people—on the Moon. The hoax spread around the world and predated Jules Verne's science fiction on the same theme by 30 years.[16]

Books and magazine articles in the United States in the 19th century were often reprints of lecture texts or propaganda for social movements and health fads.[17] Lectures by T. H. Huxley, Asa Gray, and Louis Agassiz were published in the newspapers, and in 1872 a special edition of the *New York Tribune*, which presented Tyndall's lectures, sold 50,000 copies.[18] The writers were, like these men, often leading scientists, expounding philosophical issues and the scientific worldview. They provided "not just information . . . but a model for how moral citizens should pursue life."[19] Magazines such as *Popular Science Monthly* were "evangelical publications, serving as missionaries for science among the intellectual elite."[20] Parents wanted their children to read about science, and there was a healthy market for books that offered intellectual and economic improvement through science. In one tale of scientific adventures from 1893, George Brown's father is not angry when young George uses a magnifying glass to burn a hole in a friend's coat, because "Mr Brown worked in a factory; and he was rather glad to see that his son was thinking about other things than throwing stones and kicking out his boots."[21]

At the end of the 19th century, science was a profession, and the institutional trappings that allow convenient distinctions between "scientific" and "popular" were almost in place. By then, according to historian and communications scholar Bruce Lewenstein, the "great men" of science such as Huxley and Tyndall had

> created an ideal of "public science," in which scientists deliberately used public communication about science to persuade the public that science both supported and nurtured broadly accepted social, political, and religious goals and values.[22]

In 1938 the biologist and popularizer Lancelot Hogben, summarizing the achievements of his predecessors, wrote that in the 19th century scientists had the courage and the conviction needed to communicate science because they believed that the public would understand:

In the Victorian age big men of science like Faraday, T. H. Huxley and
Tyndall did not think it beneath their dignity to write about simple
truths with the conviction that they could instruct their audience. . . .
The key to the eloquent literature which the pen of Faraday and Huxley
produced is their firm faith in the educability of mankind.[23]

## POPULAR SCIENCE FOR A NEW CENTURY

Learned societies, once accessible to anyone of appropriate social standing, were
pretty much closed institutions by the end of the 19th century. A gulf opened between the science communication that was felt to be appropriate between scientists
and the popularization that the public could appreciate. The great scientist-popularizers of the 19th century (Charles Darwin, for example), who would write their
new ideas in books accessible to a wide range of people, were replaced by the popular scientists of the early 20th century (such as the astronomer Arthur Eddington),
who would publish a journal paper and a popular book aimed at different, separate
readerships. In the 1830s and 1840s, it was possible to follow the debate about the
naming of the Devonian geological era through the pages of a popular literary
magazine such as *The Athenaeum*, which gave its readers not only closely argued
text but also detailed stratigraphical sections.[24] In the 1930s and 1940s, such
detailed discussion could be found only in the pages of the scientific journals. Popularization between disciplines became an issue early in the 20th century: a magazine called *Science Conspectus* was launched (but did not last) in the United States
in 1914 to enable scientists in one discipline to communicate with those in another.[25] The 20th century, then, sees scientific communication divided—between
disciplines within science, and between science and the public.

At the beginning of the 20th century, popular science was well established
in the United States and was in the hands of people sympathetic to science who
were proselytizing rather than responding to public demand. Their aim was the
public appreciation of science.[26] At around this time, the first targeted public information campaigns started, particularly campaigns about health, and in 1909 the
American Medical Association began a crusade against quack remedies that oc-
casioned the launch of their news bureau, in 1910. A similar campaign on behalf
of chemistry ran between the wars and was orchestrated by the American Chemical Society, which organized lobbying, press offices, and the publication and dis-
tribution of books and leaflets.

Science stories had previously been written mostly by scientists; where professional "scribes" were involved, they usually wrote down scientists' lectures ver-

batim. But as public interest in science grew, the task of reporting—in contrast to "scribing"—was falling more and more into the hands of journalists. This sometimes resulted in inventive writing: public enthusiasm for radium resulted in a number of speculative stories, including one that suggested that feeding radium to chickens could result in eggs that would boil themselves.[27] However, scientists kept a close eye on the popular press, and, when the need arose, they rushed in to set the record straight. At the turn of the century, science journalism reflected the growing division between those who felt that science was the answer to all our problems and those who felt that it might be causing them.[28] Ambivalence was polarized by war: the tanks and—most amazing of all—the airplanes demonstrated the power of science; and then there was the horror of poison gas.

Chemistry seems to offer a particular case of changing public attitudes. In 1911, Geoffrey Martin of Birkbeck College in London could write a popular book unambiguously entitled *Triumphs and Wonders of Modern Chemistry*.[29] He opened with: "The endless circulation of matter in the universe is, perhaps, the most wonderful facts [*sic*] with which chemistry has to deal." Martin's account of modern chemistry confidently took on critics from the worlds of art and literature who were antagonistic to science: "With such wonderful facts facing us on every side it is madness to assert that the progress of science means the destruction of the spirit of reverence and wonder."[30] World War I was a technological war unlike any seen before, and chemists were prominent in the war effort. "The crisis through which we and other European countries are now passing has brought home to us how greatly we, as a nation, have hitherto failed to recognize the intimate and vital dependence of our social and national prosperity on a knowledge and appreciation of the facts and principles of science, not least of chemistry, and their application in industry," commented Aberdeen chemistry professor Alexander Findlay in 1915.[31] The role of the war in forcing governments to take an active interest in chemistry was clear, Findlay said. "There are now welcome signs that the country is awakening to a sense of past deficiencies, and already the Government has taken the first step in the direction of encouraging and assisting scientific and industrial research."

Although they were not keen to be associated with some of the war's worst excesses, afterward chemists were forced to accept publicly that they did have some responsibility for the use made of their science. Nevertheless, some still managed to put a positive spin on the war effort: money and resources "on a scale before never even imagined" had been made available for research "when destruction was the sole object of humanity," E. Frankland Armstrong argued when editing a collection of works by Britain's leading chemists for the 1924 British

Empire Exhibition.[32] He added: "If the world could afford to subsidise research on the same scale, during a decade under peace conditions, our rate of progress would be beyond all belief."

But the war had led the great Swedish chemist Svante Arrhenius to other—darker—conclusions. For decades he had been warning of the profligacy of modern society and its effects on nature; Arrhenius was considering the possibility of a "greenhouse effect" and the implications of deforestation nearly a century before the 1970s Green Movement. In 1926, while remaining very positive about chemistry and its potential for social good, Arrhenius warned: "Up to the beginning of the war we lived in an age of feverish development. In ten years we used up as much good coal as constituted mankind's entire previous supply over a period of 100,000 years." By Arrhenius' calculation, coal supplies would last for another 1000 years; oil would be gone in just 20. His vision of the future for a human race that failed to take heed was bleak:

> War came and everything else was pushed aside. Now it is over and we stand contemplating the dismal ruins. We have learned by bitter experience what it means to be cut off from supplies of coals and metals. We have gained a presentiment of the desolation which may threaten our descendants in a couple of centuries in case we do not succeed very shortly in inaugurating a more sensible way of keeping house.[33]

The lesson was clear: "There holds in chemistry a rule which must be applied to all our housekeeping . . . 'Thou shalt not waste'. Herein lies our hope for the future. Priceless is that forethought which has lifted mankind from the wild beast to the high standpoint of civilized humanity."

Shortly after the war, physics captured the public imagination. Albert Einstein's theory of relativity hit the British press in 1919: *The Times* announced "Revolution in Science—New Theory of the Universe—Newton's Ideas Overthrown." Thereafter, articles about the theory were common in the press, though the opacity of many attempts to popularize it prompted *Scientific American*, in 1920, to offer $5000 to all-comers for the best popular article explaining relativity. Einstein declined to participate; and despite entries from scientists such as the Dutch astronomer Willem de Sitter, the competition was won by Lyndon Bolton, a patent clerk from London. The attention relativity attracted was variously attributed to its global significance at a time of painful international rivalries, its unsettling of British national pride, and the weirdness of bent light (we look again at this story in Chapter 6).[34] Relativity and quantum mechanics found their way into broader intellectual spheres, and thus, in a shift away from religion as a source of authority, elite or "high culture" media became forums for science.[35] On both sides of the Atlantic, scientists became public figures for their pronouncements on matters outside science: Einstein for his support for the Zionist cause; mathematician and philosopher Betrand Russell for his writings on moral issues[36]; and physicist Robert Millikan for his articles on religion.[37] Millikan's colleagues borrowed his name and invented a unit of publicity called the kan—there are one thousand millikans in every kan.[38]

The economic boom of the 1920s had positive consequences for popular science. 19th-century America enjoyed magazines; in the 1920s, science was a staple of the newspapers. From 1921, the Science Service, a national news agency for science, was distributing science news produced by and for science journalists—a step toward the professionalization of science popularization and away from the utterances of the great men.[39] The agency was a collaboration between the National Research Council, the AAAS, and newspaper magnate Edward Wyllis Scripps. Scripps had spent a long career in newspapers and had founded a number of press agencies; the Science Service was his retirement project.[40] The service, which offered "drama and romance . . . interwoven with wondrous facts" and claimed that "drama lurks in every test tube,"[41] was soon supplying copy for radio and news columns for established science magazines.[42]

In the 1920s, cinema newsreels and radio became new media for science. Einstein's visits to the United States were major events for cinema news.[43] From 1921, the U.S. Public Health Service made fortnightly radio broadcasts—a lead soon followed by the medical agencies of states and cities and by medical associations. However, business interests were a powerful lobby, and in 1931 the advice to eat less meat in hot weather prompted the introduction of censorship by the Treasury Department of health information broadcasts.[44]

The movement for the emancipation of women found some empathy with movements for the advancement of science. A new English translation of Freud's *Introductory Lectures on Psychoanalysis* was published in 1922 and had a huge impact on the young intelligentsia on both sides of the Atlantic. Freud provided scientific authority for sexual reform, as did Bertrand Russell in Britain.[45] "Scientific diets" were another phenomenon of the 1920s. In particular, "the vogue of the vitamin" prompted the American obsession with fruit juice.[46]

General-interest magazines have long been a much more prominent feature of American than of British culture. In the 1920s, publications such as *Atlantic Monthly*, *Harper's*, and *Scribner's* included features about science among the politics, arts, and fiction. Growth in popular science book publishing in the United States outstripped growth in book publishing generally.[47] The chroniclers of the 1920s recognized the public prominence of scientific culture in their fiction. In

1930, Sinclair Lewis became the first U.S. novelist to be awarded the Nobel Prize for literature.[48] Among his heroes are George F. Babbitt (1922), a real-estate broker who delights in the technological gadgetry of successful modern living, and bacteriologist Martin Arrowsmith (1925), who demonstrates the tensions of scientific life.[49] In *The Great Gatsby* (1925), F. Scott Fitzgerald's novel of American society and values, characters discuss racial theories and the idea that the Earth will eventually fall into the Sun.[50] John Dos Passos's *USA*, a historical novel of the period that was first published in 1930, includes pen portraits of Edison and of the electrical engineer Charles Steinmetz, and accounts of new scientific developments and technological consumer goods—even a drunk makes a joke about relativistic contraction.[51] With their rather narrower university education, English novelists of the period tended to ignore science: H. G. Wells and Aldous Huxley were among the exceptions. Huxley was a key figure in the intellectual life of 1920s Britain: he was a vocal supporter of reform in sexual matters, and his relatively explicit novels, in which scientists have key roles, ranted against contemporary mores.[52]

Science fiction as a genre in its modern form had grown out of the dramatic technological and intellectual developments of the late 19th century and came of age in the United States in the magazines of the 1920s.[53] It flourished in comics and on radio.[54] Karel Kapek's *RUR* brought a vision of a technological future (it was set in the 1950s) of soulless automatons and the tyranny of the machine to the London stage in 1923. With a star cast, it made as big an impact on London society as the German film *Metropolis* (1925), on a similar theme; and it preempted Huxley's *Brave New World* (1932) and Wells's *The Shape of Things to Come* (1933).[55] Science fiction would take new impetus from the new mass media of the 1950s: television and particularly film increasingly brought the images and ideas of science into popular culture. However, by the mid-1920s, science also had plenty of facts to share with the public: it was the decade of quantum mechanics, Arctic explorations, mass production, and radio, and in 1927 Charles Lindbergh flew the Atlantic—an astounding triumph for man and machine. In 1930, William L. Laurence, evangelical science reporter for the *New York Times*, wrote: "True descendants of Prometheus, the science writers should take fire from the scientific Olympus, the laboratories and universities, and bring it down to the people."[56]

The Great Depression of the 1930s brought a dip in the coverage of science, and a decline, in the United States at least, of scientists' participation in popularizing science.[57] The crash of the airship R101 in 1930 put an end to technological optimism for some: according to one English novelist, "in its brutal hideousness and terrifying monstrosity it was a symbol of what we have made of civilization."[58] However, the Great Depression also politicized the public, and many scientists were caught up in this process. Some scientists were mobilized for social and political reasons, either in celebration of the achievements and potential of science in society or, in some cases, in fear of what science, in an era of astonishing progress, might become. Socialist movements had espoused science in their publications in the United States in the 19th century; and after the Russian Revolution, and with the rise of the labor movement and universal suffrage, the public became a visible and strongly political body. From the mid-1920s, as intellectual communism gained momentum among scientists, a new motive emerged for popularizing science: rather than keeping the workers in their place, science could liberate and empower them. In Britain an influential group of scientist-socialists were also influential popularizers, of both science and socialism. In this activity, as in many other aspects of their lives, they broke with tradition, establishing new channels of communication with the public.

Preeminent among these popularizers was the population geneticist J. B. S. Haldane. Like his fellow-travelers J. D. Bernal, Hyman Levy, and Lancelot Hogben, Haldane was an unconventional figure in his professional and his personal life. But he, like the others, was also a gifted scientist and loyal to the cause of science and to the scientific community. His scientific training had taught him that popularizing was bad: in 1920, he had warned the biologist Julian Huxley, whose work had unexpectedly caught the attention of the popular press, that Huxley would "lose [his] standing as a reputable scientist, and end by being taken for a quack."[59] However, by 1923, Haldane was in full swing: his *Daedalus, or Science and the Future* was an irreverent but insightful look at such prospects as test-tube babies. According to Gary Werskey, whose *Visible College* chronicles the lives and times of Haldane and his contemporaries, Haldane's aim was "above all else, to amuse himself and others."[60] Haldane made great play, for example, of a remark by Arthur Eddington that materialists must think of their wives as differential equations:

> Men have fallen in love with statues and pictures. I find it easier to imagine a man falling in love with a differential equation, and I am inclined to think that some mathematicians have done so. Even in a non-mathematician like myself, some differential equations evoke fairly violent physical sensations similar to those described by Sappho and Catallus when viewing their mistresses. Personally, I obtain an even greater "kick" from finite difference equations, which are perhaps more like those which an up-to-date materialist would use to describe human behaviour.[61]

But whatever Haldane's talent to amuse, his politics were never out of sight. He frequently wrote science articles for the communist newspaper the *Daily Worker*. He urged that "a knowledge of science should be spread among socialists, and a knowledge of socialism should be spread among scientists,"[62] and many of his essays are primarily expository and didactic. By 1946, the man of experience was ready to write an essay entitled "How to Write a Popular Scientific Article." Haldane stressed that his advice was merely his personal opinion—he remarked, for example, with reference to overcomplicated sentences, that "Prof. Hogben writes sentences longer than some of my paragraphs, and his books sell very well, as indeed they ought to."[63]

Lancelot Hogben, like Haldane, was aware that popularizing could incur the wrath of other scientists. Their view, he remarked in *Science for the Citizen* (1938), was that a scientist who writes a book that "can be read painlessly . . . neglects his students, neglects his laboratory, and neglects his golf."[64] Hogben took this attitude seriously: his *Mathematics for the Million*, which became an international best-seller, was due to go to press at a time when Hogben felt he stood a good chance of being elected a Fellow of the Royal Society, and, rather than jeopardize his FRS, he asked his colleague Hyman Levy to pretend to be the author of the book. Levy declined, and the book was published under Hogben's name, but only after the honor had been confirmed.

Socialist groups encouraged their scientist-members to stick to the rules of science: as public socialists, they had also to be good scientists, both in the laboratory and within the social structure of the scientific community. The scientific establishment saw it rather differently: any political stance could only detract from the pursuit of scientific excellence. Hogben's friend the embryologist and historian of science Joseph Needham used a pseudonym for his book on the English Civil War and even reviewed other books on that subject under his assumed name. But in essence the scientists' role was, according to Werskey, "to serve as socialist spokesmen for the culture of science," and the scientific culture was well served by them. Their activities opened up new routes for popular science in Britain, and the scientist as opinion leader become commonplace. Radio broadcasts, book series, essays, and encyclopedias would aim to equip the public for citizenship in a scientific age.

The spirit of the time is exemplified by H. G. Wells, Julian Huxley and G. P. Wells—H. G.'s son "Gip"—in their popular classic *The Science of Life*, which was published in 1931. Wells Sr. had no truck with the idea that popularizing was a minor activity: he stressed to his co-authors that "*this job is an important job*; your own researches and your professional career are *less important."*[65] They had a broad vision of what scientific knowledge could do for "the ordinary man":

> A great and growing volume of fact about life as it goes on about us and within us becomes available for practical application. It reflects upon the conduct of our lives throughout; it throws new light upon our judgements; it suggests fresh methods of human co-operation, imposes novel conceptions of service, and opens new possibilities and new freedoms to us.[66]

But the busy layperson faces many problems in accessing this knowledge, which is dispersed in numerous books and often obscured by technical language. Also, the knowledge may not be reliable, for

> it is mixed up with masses of controversial matter, and with unsound and pretentious publications. Complicated with the clear statements of indubitable science there is a more questionable literature of less strictly observed psychology, and a great undigested accumulation of reports and statements turning upon current necromancy, thought transference, speculations about immortality, will-power and the like. Mingled with orthodox physiological teachings are the doctrines of dietic dogmatists, and the prohibitions and injunctions of religions and other regimens. Obscuring the facts of heredity are the heavy accumulations of prejudice and superstition.

Because of this:

> In the care of his health and the conduct of his life, the ordinary man, therefore, draws far less confidently on the resources of science than he might do. He is unavoidably ignorant of much that is established, and reasonably suspicious of much that he hears. He seems to need . . . the clearing up and simplifying of the science of life . . . .

The authors themselves claim innocence of the bias they detect in others' writings: they are "wedded to no creed, associated with no propaganda; . . . telling what [we] believe to be the truth, so far as it is known now." Despite that, their motives are, at least in part, political:

> Antagonism to biological knowledge is by no means dead. There is a constant struggle to keep physiological or pathological information from people who might put it into beneficial practice, and to prevent the complete discussion of such questions, for example, as the possible control of the pressure of the population upon the reserves of the community. There is little or no reasoned justification for these suppressions. In some of the more backward regions of the United States, moreover, there is a formidable campaign for the penalization of any

biological teaching that may run counter to the literal interpretation of the Bible.

Socialism may have been a force for popularization in Britain, but in the United States, capitalism was hard at work. In 1933, the U.S. President's Research Committee on Social Trends announced the coming of age of a new type of American: the consumer—someone who buys the material of their daily life, instead of gathering, making, or growing it. For example, "The homely broom that had remained unchanged since the time of the early Egyptians is giving way to an expensive piece of electrical equipment."[67] Science and technology were cited as key factors in this change, and concern was expressed about consumer literacy, in science and in other areas such as safety and budgeting. Public schools were beginning to teach diet and health, though still mostly only to girls, and the government extended its interests in the standards and safety of consumer goods and the truthfulness of advertisements. Commercial bodies such as the National Dairy Council, the American Gas Association, and the Kellogg Company encouraged public awareness of their products with demonstrations, informative commercials, and news bulletins for teachers. Life insurance companies organized extensive programs of public health information, designed to help people live happier, longer, more productive lives in the workplace and at home, which included publications and a "visiting nurse" scheme. (These happier, healthier people would also, of course, pay more in their longer lives to their insurance agent.) For working people with limited access to formal education and the mass media, their insurance agent was often their only source of reliable health information.[68]

## SCIENCE IN POLITICS

During World War II, science became an integral part of the public information effort in Britain, particularly with regard to health and nutrition. A new voice emerged: radio became the first truly mass medium in Britain, uniting the nation in a time of crisis. From 1942, on the BBC's Home Service, the Radio Doctor gave advice on diet and health to people struggling with food rationing and limited medical services.[69] When the men went off to war, women scientists became a feature of U.S. science coverage, both as subjects and as authors of magazine articles.[70]

However, the popularization of frontier science all but ceased during World War II; scientists' work was often secret, and publishers and paper were put to other uses. Scientists could still communicate the value of science, and their mysterious war work gave them a new heroic image. In an anthology of science writing published in 1943, its editor, the distinguished Harvard astronomer Harlow Shapley, described popular science in romantic terms:

> Perhaps the greatest satisfaction in reading of scientific exploits and participating, with active imagination, in the dull chores, the brave syntheses, the hard-won triumphs of scientific work, lies in the realization that ours is not an unrepeatable experience. Tomorrow night we can go out again among the distant stars. Again we can drop cautiously below the ocean surface to observe the unbelievable forms that inhabit those salty regions of high pressure and dim illumination. Again we can assemble the myriad molecules into new combinations, weave them into magic carpets that take us into strange lands of beneficent drugs and of new fabrics and utensils designed to enrich the process of everyday living . . . . We can return another day to these shores, and once more embark for travels over ancient or modern seas in quest of half-known lands—go forth as dauntless conquistadores, outfitted with maps and gear provided through the work of centuries of scientific adventures. But we have done enough for this day. We have much to dream about. Our appetites may have betrayed our ability to assimilate. The fare has been irresistibly palatable. It is time to disconnect the magic threads; time to wind up the spiral galaxies, roll up the Milky Way and lay it aside until tomorrow.[71]

That introductory essay is dated 1943; Shapley's anthology was reissued in 1946 with a new section on the atomic bomb. Before the United States joined the war, the power of the atom had been trumpeted from the pages of the *Saturday Evening Post*, and public expectations had been high. In the new section, Shapley argued that, however horrendous the potential of this new weapon, no one would doubt "the ultimate beneficence of increasing knowledge." Journalist John O'Neill speculated that atomic-powered cars could have a lifetime's fuel built into them on the assembly line and that electricity would be so cheap that sidewalks would be electrically heated to save us the chore of shoveling snow. But the last word of the 1946 edition went to atomic physicist J. Robert Oppenheimer, who had directed the Manhattan Project to build the first atomic bomb. He cited a colleague who cautioned him against discussing openly the horrors of atomic warfare, because it would turn the public against science. Oppenheimer took the opposite stance:

> I think that it is for us among all men, for us as scientists perhaps in greater measure because it is our tradition to accept and recognize the strange and the new, I think it is for us to accept as fact this new terror, and to accept with it the necessity for those transformations in the world which will make it possible to integrate these developments into

> human life. I think we cannot in the long term protect science against
> this threat to its spirit and this reproach to its issue unless we recognize
> the threat and the reproach . . . . Whatever the individual motivation
> and belief of the scientist, without the recognition from his fellow men
> of the value of his work, in the long term science will perish.

Oppenheimer was a popular hero and a respected public figure in the United States immediately after the war. According to sociologist Philip Reiff, "Oppenheimer became a symbol of the new status of science in American society. His thin handsome face and figure replaced Einstein's as the public image of genius."[72] Oppenheimer's later arguments against nuclear weapons development were, like those of many of his former colleagues on the Manhattan Project, largely moral arguments. But he made them in public. Oppenheimer's security clearance was revoked in 1953, and in testimony at the subsequent hearings on this decision, the atomic physicist Edward Teller said: "I would feel personally more secure if public matters would rest in other hands."[73] Oppenheimer's security clearance was not reinstated.

In 1940, a group of British scientists had met over dinner to discuss the war. A month later a book called *Science in War* was on station bookstalls across Britain. The scientists had left the dinner determined to present their arguments for the effective use of science in the war effort to as wide a public as possible and had persuaded Penguin Books to publish their essays. The book was published anonymously; J. D. Bernal and Solly Zuckerman had been among the diners. *Science in War* claimed:

> Science is the most orderly expression of normal ways of acting and
> thinking. There is therefore a great need today for quick scientific
> thought. Only the scientific method can deal effectively with the new
> problems which turn up daily, and the issue of the war depends largely
> on how quickly and how effectively science is used. . . . It is essential
> to break down the traditional resistance of administrators to a general
> scientific approach to the problems of national existence.[74]

As in the social upheavals of the 1930s, many scientists had become politicized by the war; organizations such as the British Society for Social Responsibility in Science were set up by scientists concerned, for example, by the enormous power of atomic weapons and other wartime inventions. And because many academic scientists were taken out of the universities and put to work for government or military agencies, the scientific community had an inside look at the seats of national power. Some of them were less than impressed by what they had seen. One such was the British cosmologist Fred Hoyle, who had spent the war working for the Admiralty on radar. In 1950 he was voted Broadcaster of the Year by BBC listeners: his radio talks, "The Nature of the Universe," had been an extraordinary success, and Hoyle became a household name. Hoyle had become impatient with the state apparatus—he was "genuinely puzzled by what had gone wrong" in postwar Britain.[75] By 1953 he was ready to launch his own manifesto: his book *A Decade of Decision* presented solutions that, he believed, would be more effective than those the politicians were trying to foist on an ill-informed electorate. The book was dedicated "to those who are determined that Britain shall not decline into apathetic dependency." Hoyle claimed that his motive was his duty to his country, and he recognized that society does not usually turn to scientists for solutions to social problems:

> [When] scientists . . . keep silent on social and political matters, they are accused of being indifferent to the abuse of their discoveries. When they speak out, they are usually told that their views are too "calculated", or that they are too "naive", or that they are guilty of "oversimplifying difficult problems" . . . . This has not dissuaded me . . . because the successes . . . achieved by the people who claim to understand the subtlety of human problems have not been impressive. This being so, I do not think that the non-expert need be unduly bashful about putting forward a new outlook on human affairs.[76]

*A Decade of Decision* was widely reviewed. The *Daily Telegraph* wrote: "The distinguished Cambridge astronomer's excursion back to earth is not very helpful from a practical standpoint, but the book is eminently worth reading for its lucid statement of a serious problem."[77] The *Economist* magazine was less complimentary:

> When a man of outstanding intellectual attainments uses them to cultivate a field in which he has no special expertise the results may be illuminating or plain silly. Mr Hoyle's *Decade of Decision* is both . . . one may wonder at the naivety which allows him to ignore . . . everything that is organic in the relations of interacting economic units; all considerations of scale, specialization and indivisibility; all that makes the difference between the tabula rasa of his imaginary empty Britain and the infinitely complex palimpsest of a modern industrial society.[78]

## THE POST-WAR BONANZA

The end of World War II heralded an outpouring of scientific information: the war effort had produced new technologies in medicine, energy generation, transport, and communications. Newspapers took up the coverage of science again: the morale-boosting, optimistic rhetoric that carried readers through the war persisted, and the accelerated technological achievements of the war years were celebrated in similarly upbeat terms.[79] Attitudes to science were generally positive: Winston Churchill claimed that without science the war might not have been won.[80] The late 1940s and in particular the 1950s were celebratory times for science, and new mass media brought opportunities for the scientist as both celebrant and celebrity.[81] But during the the war, the conflicting responsibilities of secrecy and public information had divided scientists into two camps: those whose duty it was to keep quiet, and those whose duty it was to be in the public domain. Both groups—those used to secrecy and those for whom popularization meant public information or propaganda—were ill at ease in the new post-war climate. Increasingly, journalists took over as the communicators of science, and scientists became sources at one remove from their audiences.

Post-war popular science was much more organized than its pre-war counterpart. Particular professional groups, each with its own agenda, mission, and strategy, took over from the dabblers, idealists, and stars of the 1920s and 1930s. Lewenstein has identified four such groups that emerged in the United States in the late 1940s: scientific organizations, commercial publishers, the government, and the science writers. All of these groups had been active before the war, and the war itself presented each with a different set of challenges and opportunities. But the group that experienced the biggest change at this time was the science writers—indeed, the post-war period in popular science is perhaps characterized by the rise and rise of the science journalist. For while journalists have been reporting science in the British media since the early 19th century, specialist science media professionals are a feature of recent British history. Science journalist Ritchie Calder claims to have been one of only three specialist science journalists in Britain in the the 1930s[82]; by 1947, there were enough to merit the founding of the Association of British Science Writers. In the United States, the National Association of Science Writers (NASW), which had been founded in the early 1930s, had 63 members in 1945, but twice that many by 1950, and 413 members by 1960. While the NASW was committed to unbiased reporting, and marginalized those science journalists who took up public relations posts with particular scientific institutions, its members were, according to Lewenstein, "committed to promoting science as the saviour of the world" and viewed themselves as "advocates for science."[83]

President Truman's speech announcing the dropping of the atomic bomb had been written by William L. Laurence, the self-proclaimed Promethean who had defined science journalism in the United States in the 1930s and who had been invited in 1945 to observe the Manhattan Project. Laurence had warned his editor to expect a scoop comparable in magnitude to the second coming of Christ.[84] Stephen White, reporter for the *New York Herald Tribune*, recalled in an article for *Physics Today* in 1948 that his career as a science journalist had begun on August 6, 1945. White's article provides an account of the problems that arose as the communities involved in science communication became more specialized and distinct. His concerns were for the newspaper-reading public, who do not understand science, and for the physicists, who do not understand journalism, and for himself, caught somewhere in between. White advised physicists to judge the reporter by his paper and to check whether the reporter was a member of the NASW rather than assuming he was irresponsible and ignorant; he suggested that physicists give background information in advance so that reporters had time to prepare for the big news story. He asked physicists to try to understand the news process:

> Whatever else may be true of our writing, it is generally adapted to the purposes of our newspapers. If it is rich with cliches, they are *our* cliches, and we use them because they enable our truck drivers to get the gist of a paragraph without trundling out the dictionary. . . . A news account, by its very nature, is a compromise between the facts and the general impression. Beyond a certain point, what it gains in precision it loses in communication. [A technical word] obscures a news story because it . . . hides the meaning for most readers. . . . The terminology of the layman is an absence of terminology; the precision of the layman is an accuracy of impression rather than an accuracy of specific fact; newspapermen, like physicists, vary and must be approached according to their reliability; and newspapermen, like physicists, have certain over-all requirements they must fulfill, and physicists, in dealing with newspapermen, must be guided by those needs.[85]

These requests have been made by journalists of scientists many times in the years since White wrote this article. Nevertheless, both parties seem to have survived the scientist–journalist interaction: science was increasingly part of national economies and of national and international politics in the 1950s and became an established part of news coverage. The late 1950s and early 1960s were a boom time for science coverage, and for science on the front page; in the late 1950s in particular, science articles were awarded a prominent position in the newspapers.[86] The physical sciences were for many years the most prominent sciences in the British quality press, and it is only since the mid-1970s that they have lost their lead in the popular press to the biomedical sciences.

As we noted in Chapter 1, the launch of Sputnik in 1957 astonished the world. That the Russians should have won the space race was unthinkable: the reaction in the United States in particular was close to panic. President Eisenhower responded to journalists' questions about U.S. science with the remark, "Well, let's get this straight. I'm not a scientist"; and scientists leaped into the public arena armed to the teeth with ideas for putting the United States back in its rightful place at the top of the scientific heap.[87] The media stopped bemoaning the backwardness and vulnerability of the United States and started hyping American science. Scientific institutions and lobby groups marched into the public domain trumpeting science. A good example of the "Sputnik effect" is *An Inexpensive Science Library*, a list of recommended reading first compiled in 1957 and published annually by the American Association for the Advancement of Science. By 1963 the list had become a book: *A Guide to Science Reading* is dedicated to "the scientists of tomorrow," and consists of some scene-setting essays, followed by a list of titles and brief outlines of books that can provide the citizen with a grounding in science. The foreword, by Hilary J. Deason of the AAAS, pulls no punches:

> Scientific literacy has become a real and urgent matter for the informed citizen. Many have allowed themselves to lapse into a coma of scientific illiteracy because of the misconception that science and mathematics are beyond their understanding, have no personal appeal, and can be rejected or ignored. For them a tocsin has been sounded by hundreds of intelligent men and women whose personal crusade is the awakening of people everywhere to scientific awareness. They admonish, "read, mark, learn, and inwardly digest." They have spread an intellectual feast so rich and varied that scientific illiterates find that sharing in it is an exciting experience.[88]

The scientific illiterates not too shamed by their own inadequacies to venture past the foreword would have found 177 books recommended in the 1957 edition—a list that had expanded to over 900 entries by 1964. The authors ranged through history from Galileo to Margaret Mead and were from throughout the English-speaking world. All the big names are there—except, of course, for Oppenheimer and the English socialists. The introductory essay by Warren Weaver, entitled "Science and People," is quoted in Chapter 1; Margaret Mead's essay echoed Oppenheimer's claim for openness; and H. Bentley Glass, Professor of Biology at Johns Hopkins University, concluded his essay with the following italicized exhortation:

> *. . . the general citizen of his country, the man in the street, must learn what science is, not just what it can bring about. Surely this is our primary task. If we fail in this, then within a brief period of years we may expect either nuclear devastation or worldwide tyranny. It is not safe*

*for apes to play with atoms.* Neither can men who have relinquished
their birthright of scientific knowledge expect to rule themselves.

Glass was a teacher, and his essay stressed the challenges to which teachers of science had to rise. Indeed, perhaps the most lasting Sputnik effect was the reorganization of science education, particularly in the United States: during the 1960s, the number of people who learned science at school increased enormously, and these citizens were as well equipped to criticize as to support science. The protest movements of the late 1960s and early 1970s, particularly the environmental and antinuclear groups, were driven by young people whose political activity was informed by their newly acquired scientific knowledge.[89]

Key popularizers of this period were not lecturers or writers: they were broadcasters. Although scientists and science writers achieved commercial success and popular acclaim with books and articles, their readerships were tiny compared to the audiences for science broadcasts. After a brief preeminence for radio immediately after the war, television took over as the mass medium for science. The TV scientists—Jacob Bronowski and Jacques Cousteau—and their colleagues the TV hosts—Patrick Moore, David Attenborough, and Walter Cronkite—personified science in public.

The history of science on television has yet to be documented. However, the development of science programming at the BBC perhaps offers some insights into the changing fortunes of popular science in the post-war years. Science appeared early in the history of television: the BBC's first science program was the short-lived "Inventor's Club," which was first broadcast in April 1948. The BBC's listings magazine *Radio Times* announced "a promising and important new project . . . to find the latent inventive genius in this country."[90] The program consisted of a panel of experts, drawn from science and industry, who examined devices made by members of the public. The producer hoped that ex-soldiers might put their military training to use in inventions that would boost the exhausted post-war economy. Sadly, the expert panel had few encouraging words for the aspiring inventors, and the program made uncomfortable viewing.

"Science Review," screened in February 1952, was the first full-length science documentary; it was announced as "a film review of recent research and discovery in science and industry" and was watched by 4 million people—over 10 percent of the British population. A year later came "The Quatermass Experiment," the BBC's first drama series about science and scientists. The first adventure concerned a space mission that went wrong, and was watched avidly by 5 million people. The program was a symptom of and response to the contemporary obsession with science fiction. It captured the mood of uncertainty about science and the future after the first Soviet atomic bomb test in 1949 and the launch of the American H-bomb program in 1950.

"Zoo Quest"—a combination of earnest science and real-life safari, catching animals for British zoos—was first shown in 1954. One of the BBC's representatives on the expedition was David Attenborough, who was to become the driving force and guiding light for the presentation and interpretation of natural history on television. However, in this first program, the BBC's role was not analysis or comment but simply "to record the progress of the expedition, the ways in which the animals were collected, our meetings with African chiefs and hunters, and to film animals of all kinds in their wild state." The immense popularity of "Zoo Quest"—20 percent of the British population watched it—was an indication of the successes to come for natural history on television around the world.

The BBC's astronomy show "The Sky at Night" was broadcast for the first time in April 1957. The night skies that year boasted two comets that could be seen with the naked eye, as well as Sputnik, and "The Sky at Night" was watched by 10 percent of the population. However, the BBC was evidently nervous about its subject matter. The *Radio Times* thought it wise to explain to potential viewers that astronomy need not be quite as esoteric as they might think:

> Stars, in our particular workaday world, tend to be people with agents, and success stories, and temperaments, and talents, and a certain unpredictability that can make them either tiresome or fascinating. But when we talked to [the host] Patrick Moore the other day about this month's stars, we found ourselves in a very different world. People tend to think, he told us, that astronomy is a difficult, expensive and unrewarding subject that has become the prerogative of old men with long white beards. It is in fact none of these things, and anyone can find interest in the night sky, if he knows what to look for.

"The Sky at Night" is still broadcast today and is still presented by Patrick Moore. It is the longest-running program in the history of British television.

"Your Life in Their Hands" was a breakthrough. Starting in 1958, a BBC crew took their cameras into hospitals and showed the medical staff at work. They were given permission to do this by the Ministry of Health after they had undertaken "only to interpret the facts and not to impose a false sense of drama for the sake of sensationalism." Another new feature was that the program was presented by the doctors themselves:

. . . there are no actors or professional commentators. Physicians and surgeons, with the help of every kind of visual aid, will explain and demonstrate. For most of them it will be a new experience.

The BBC saw "Your Life in Their Hands" as a learning experience for the viewer. The producer Bill Duncalf wrote that knowledge of the circumstances and conditions under which medical science was practiced was vital if we were to appreciate the advances in medical knowledge, which were usually presented as facts isolated from their consequences. As well as learning about the processes of medicine, the "viewer is offered the opportunity of learning from the most eminent practitioners something about his body." "Your Life in Their Hands," narrated straight from the mouths of specialists, also presented the producers with the problem of jargon. As Duncalf explained at the time, "the main problem has been the translation of specialist language into lay terms and it will require some concentration on the part of the viewer if we are to achieve success, but it should be within the capabilities of everyone to understand what is happening."

The most enduring science documentary series has been "Horizon," which was first broadcast in 1964. Its audience ratings vary with its subject matter, but it can claim up to 5 million viewers in Britain. The programs are held in high regard and are broadcast throughout the world, particularly in partnership with the U.S. series "NOVA." The magazine "Tomorrow's World" was first broadcast in July 1965 and consisted of "film, outside broadcasts and studio reports on the men and developments which are changing our way of looking at and living life." Its commitment to contextualizing its science has wavered in the intervening years, and the program has received some criticism for its "gee-whiz" approach; it has, however, held on to its prime-time slot for over 30 years. With its trickery and gadgets, "Tomorrow's World" can claim to have given many British children a first glimpse of the excitement to be had from science.

Much of the BBC's television science became the models for the rest of the world. The United States, though, has been innovative in many ways. One opportunity was provided by the extensive public subscription to cable television, which has provided an audience big enough to sustain an entire channel devoted to science. The Discovery Channel offers a rich diet of programs of different formats and on a wide range of sciences, making readily available the documentaries and magazines that fight for airtime in the crowded schedules of mainstream television channels in other countries. The Discovery Channel and other channels are increasingly involved, too, in large-scale international coproductions of popular science. Another achievement of U.S. television is its children's programs about science. In

"Mr. Wizard's World" on MTV, children who drop by Mr. Wizard's house do experiments with scientific apparatus and ask Mr. Wizard to explain what they see. In "Beakman's World" on CBS, Beakman, his assistant, and a pantomime rat do experiments to demonstrate the answers to viewers' scientific questions. "Bill Nye the Science Guy" is broadcast on PBS from "Nye Labs," where Nye, a stand-up comedian (and former mechanical engineer), uses experiments, location reports, and conversations with scientists to examine a particular scientific theme.[91]

The close connections between science popularization and education in the United States, and the commitment of scientific and charitable foundations to the public understanding of science as they see it, has created an environment in which groups with a particular agenda can broadcast their own message about science. Television is, for example, a medium that has been used by particular agencies for health education and dietary advice, delivered in the form of documentaries. Another example is the series "Discovering Women," which was made, according to one review, "as part of a national educational program to change the way middle-school students approach science." The programs were broadcast on PBS and were funded by the National Science Foundation, the Alfred P. Sloan Foundation, and the Intel Corporation. Their explicit aim was to "enhance public understanding of the contributions made by women in science."[92]

However, television has not always promoted a pro-science message. British television science took up the watchdog role in 1970, though only tentatively and through drama. Gerry Davis, a scriptwriter, who had cut his teeth on the classic science fiction series "Dr Who," teamed up with a scientist and wrote "Doomwatch," a series about the dangers of unscrupulous scientific research. Described as a "fictional drama series, frighteningly close to reality," it showed the work of a group whose job was to monitor and prevent abuses of the environment or people by scientists and science-based industry seeking money or power. Davis wrote in the *Radio Times*:

> The days when you and I marvelled at the miracles of science—and writers made fortunes out of science fiction—are over. We've grown up now—and we're frightened. The findings of science are still marvellous, but now the time has come to stop dreaming up science-fiction about them, and write what we call 'sci-fact'. The honeymoon of science is over . . . .

When the environmental concerns of the early 1970s raised questions about science, the BBC became bolder and extended its critical position from drama to documentaries. "Horizon" stopped being simply reportage and became investigative journalism and was influential in changing attitudes and, occasionally, laws.

Programs on whaling, asbestos, and alternative energy had a profound impact on growing audiences.

As on television, so in the press. The new environmental science appeared often in the newspapers, but much of the coverage was critical of technology and the other sciences. Since the 1960s, science journalists had seen themselves less as missionaries for science and more as critics and commentators—like their colleagues on other beats. In the United States, John Lear of the *Saturday Review* wrote that "the spirit of untrammelled inquiry and scepticism required of journalism in other fields must become a standard in science writing."[93] Indeed, until the early 1960s, science in the mass media had been generally positive in tone[94]—scientists themselves and general reporters and broadcasters had communicated a pro-science message. The change to a more critical voice during the 1960s was partly a response to the disappointment that followed the pro-science euphoria of the immediate post-war period and partly a response to perceived failures of technology, especially with regard to energy and the environment. It was also due to the emergence of a new breed of science-trained science journalists who engaged critically with both science and its issues. This increasingly aggressive journalism in the 1960s, and the rise of the celebrity science journalist, such as John Maddox of the *Manchester Guardian* and Walter Sullivan of the *New York Times*, further alienated scientists from direct contact with the public. Communications researcher Christopher Dornan reports a view prominent in the 1960s of science journalists as "exploiting science as a source of startling narratives."[95] For a scientist, then, to popularize science was to exploit his or her own community. By the 1970s, popularization was an activity in which all but a few high-profile "visible" scientists engaged at considerable professional risk (see Chapter 4).[96]

Once again, then, scientists could participate in public culture only at the expense of their own participation in the scientific culture. As we saw in Chapter 1, the broader implications of the scientists' withdrawal into their own community were soon felt, and arguments about the interdependence of science and society emerged once again. The relative frenzy of popularization during the 1980s and 1990s and its legacy at the end of the 20th century are discussed further in Chapters 9 and 10.


# THE TWO CULTURES ... AND A THIRD?

Despite centuries of science popularization, despite decades of history in the collaboration between science, government, the media, and the public, despite the apparent increase in the amount of science in public culture, and despite the growing numbers of people with some basic education in science—despite all this, the role of science in public remains a matter for argument and ambivalence. One recurring debate concerns the relationship between science and culture as a whole, and between science and the arts, in particular. Spokespeople for science have periodically complained that, at least as far as the Anglo-Saxon tradition is concerned, science is not recognized as a worthy component of social culture. Science, they feel, is less highly regarded in the popular imagination than art, literature, or music. In part, it is argued, this is because those who control the outlets of contemporary culture have been trained to be indifferent, if not actively hostile, to science. Put crudely, for example, it is claimed that science journalists cannot get their stories into the mass media because all their editors are arts graduates with a blind spot for science.[97] This perception of the place of science in culture—that there are, in effect, two cultures: science and the rest—has become part of popular wisdom. According to historian of popular science Marcel LaFollette, "we routinely accept as a fashionable truism the argument that there are 'two cultures'."[98]

Although C. P. Snow was far from being the first to remark on it, the key statement of the "two cultures" thesis is due to him. A science administrator, political novelist, and one-time scientist, Snow became one of the most provocative commentators on culture in Britain in the 1950s and 1960s. He first drew attention to what he saw as a growing cultural divide in 1956, in an article in the left-of-center magazine *New Statesman and Nation*. The version of Snow's argument that is now most widely read was given as the Rede Lecture of 1959 in the Senate House of Cambridge University.[99] "I believe," Snow said, "the intellectual life of the whole of western society is increasingly being split into two polar groups."

In giving his talk, Snow drew heavily on his own career. Describing his intellectual peregrinations, he explained:

> I felt I was moving among two groups . . . who had almost ceased to communicate at all . . . instead of going from Burlington House [then home of the Royal Society of London] or South Kensington [Science Museum] to Chelsea [famed as the artistic quarter of London], one might have crossed an ocean . . . . Literary intellectuals are at one pole—at the other scientists . . . . Between the two a gulf of mutual incomprehension—sometimes . . . hostility and dislike, but most of all, lack of understanding . . . . The separation between the scientists and the non-scientists is much less bridgeable among the young than it was even 30 years ago . . . at least they managed a frozen smile across the gulf. Now the politeness is gone and they just make faces.

Although, at the beginning of his lecture, Snow adopted a cultural neutrality—he chided scientists for their "shallow optimism" and "cultural self-impoverishment"—it soon became clear that what he termed the "traditional culture," which he broadly identified with "literary culture," was the problem as far as he was concerned. The accusation was that this literary culture was behaving like a state whose power was rapidly declining but which was too concerned with defending itself to recognize the very forces that could help in reshaping it. Literary intellectuals were "natural luddites," who looked on the centuries of agricultural grind and poverty as if they were a golden age; their unconcern with the real state of humanity had led them to support Mussolini and Hitler, Snow accused. Scientists, on the other hand, had "the future in their bones."

Snow attributed the problem of the cultural divide, in Britain at least, to the early choices school students had to make; at the age of 14, many had to give up either science or languages and the arts, and specialize. But the two-cultures phenomenon was not solely a British (or, even, English) problem, Snow felt; it affected the entire Western world. The consequences too were worldwide: the widening gap between rich and poor nations could only be addressed if scientists were dispatched to all corners of the globe to muck in and help with technical and economic development. The failure of Western rulers, steeped in the traditional culture, to recognize the importance of science and scientists in this global evangelizing role would lead to the gap between rich and poor widening, with disastrous consequences. "We have little time," Snow warned.

Snow's thesis was very much a product of the man and his times. In 1956 Britain had been humiliated when it had tried to wrest back the Suez Canal from the emerging nationalist regime in Egypt. The failure to do so signified a decline in Britain's imperial power. There was a musty air about the British ruling elite that many identified with its Oxbridge classics training.

Snow himself had begun his intellectual life as a scientist and had then moved into the administrative side of research in both government service and private enterprise. When his research career hit difficulties, he began another career as a novelist, specializing in plots involving clashes between scientists and politicians. By the time he came to give the Rede Lecture, he also had two decades' experience of administering science and had achieved considerable stature as a statesman. Snow therefore felt he knew intimately both the cultures he identified: he felt his statesmanship made him eminently qualified to discuss these large matters, and his authority rested on his own wide-ranging personal experience. So it was inevitable that critics would challenge Snow on his qualifications as cultural theorist of the post-war world. The most famous attack on the concept of the two cultures was markedly *ad hominem*. The literary critic F. R. Leavis insulted Snow the novelist and claimed that the fact that his thesis had achieved such widespread acceptance was in itself a sign of the times:

> Snow is a portent. His significance is that . . . he has been created an authoritative intellect by cultural conditions manifested in his acceptance . . . . Snow's relation to the age . . . is characterized . . . not by insight and spiritual energy, but by blindness, unconsciousness and automatism. He doesn't know what he means and he doesn't know he doesn't know.[100]

But perhaps Leavis' most telling argument was that, given the rapidity of the change that Snow's beloved science and technology were bringing to the world, mankind would "need to be in full intelligent possession of its full humanity."

> What we shall need, and shall continue to need not less, is something with the livingness of the deepest vital instinct; as intelligence, a power . . . of creative response to the new challenges of time; something that is alien to either of Snow's cultures.

Ever since Snow pointed up the "two cultures" divide, historians have looked for its origins and sociologists have investigated its mechanisms and its consequences. Snow himself identified the traditional literary culture with the Romantic reaction to science prevalent at the end of the 18th century, when science was becoming more professional. However, historian Stephen Toulmin considers the cultural differences between the arts and sciences, such as they are, to date as far back as the Renaissance.[101] The arts and humanities, he argues, had undergone their fundamental, modernizing, changes by the end of the 16th century. This was an era of relative political stability in which liberal tolerance and subjective humanity could be incorporated into the culture. In comparison, the 17th century, the period in which science was modernized, was one of turmoil. As a result, science developed as a disciplined, objective, and deliberately unhumanized part of modern culture, a sure and certain haven from the political mayhem around it. The "two cultures" took on separate characteristics almost from the outset of their modern forms, according to Toulmin.

The American literary critic Lionel Trilling saw the debate between Snow and F. R. Leavis as a reprise of that between T. H. Huxley and the poet and literary intellectual Matthew Arnold that took place in the 1880s, and which was closely followed on both sides of the Atlantic.[102] In 1880, Huxley was asked to open a college in Birmingham that was to be funded by Sir Josiah Mason, a wealthy industrialist. Within the strict prescription for Mason's College was that "the College shall make no provision for mere literary instruction and education." This injunction chimed with "two convictions" that Huxley held strongly about education:

> The first is, that neither the discipline nor the subject matter of classical education is of such direct value to the student of physical science as to justify the expenditure of time on either; the second is, that for the purpose of attaining real culture, an exclusively scientific education is at least as effectual as an exclusively literary education.[103]

Society would benefit from more scientific education: it would allow young artisans and their future employers to "sojourn together for a while," learning that "social phenomena" (such as their respective positions in society) were "as much the expression of natural laws as any other." Huxley realized that his opinions were out of line with those of most of his contemporaries but saw himself continuing a tradition in the battle for education:

> In the last century, the combatants were the champions of ancient literature on the one side, and those of modern literature on the other; but some thirty years ago, the contest became complicated by the appearance of a third army, ranged round the banner of Physical Science.

And now, Huxley argued:

> . . . an army, without weapons of precision and no particular base of operations, might more hopefully enter upon a campaign on the Rhine, than a man devoid of a knowledge of what physical science has done in the last century, upon a criticism of life.

But, Huxley complained, the classical scholars who dominated the educational science would not accept that science could "confer culture" on its students; they could not, the classicists said, be accepted into "the cultured class."

Huxley had in mind the likes of Arnold, the self-appointed culture guru of the 1880s. Arnold, whose definition of a cultured person was someone who knew the best that had been thought and written, accepted science as part of culture insofar as it consisted of written text. But the knowledge that science gave was merely one damned thing after another—"knowledge NOT put for us into relation with our sense for conduct, our sense for beauty . . . and therefore . . . unsatisfying, wearying."[104] Confidently, Arnold declared:

> If then there is to be separation and option between humane letters on the one hand, and the natural sciences on the other, the great majority of mankind . . . would do well . . . to choose . . . letters. I cannot really think that humane letters are in much actual danger of being thrust out

> from their leading place . . . so long as human nature is what it is, their
> attractions will remain irresistible.

Whatever its true historical antecedents, C. P. Snow's two-cultures thesis received both enthusiastic support and fierce criticism from the moment it was published. From one side, he was hailed as having identified a deep social and historical problem. From the other, Snow was attacked for his dialectical use of the word "two" and his ambiguity about what he meant by "culture." (He quipped that no one had actually quibbled with the definite article.) Historian Daniel Kevles claims that Snow was, to the interacting cultures of science and politics in the United States in the mid-1960s, a "high priest."[105] But as Stefan Collini has recently pointed out, much has happened in the four decades since Snow was writing and lecturing.[106] Changes to the British education system, for example, now preclude students dropping all science at an early age, and television brings science and natural history programs into the home as part of "mass culture." Professor of English George Levine and co-thinkers are among many who go still further and argue against seeing any chronic separation between science and the arts, pointing to the continuous cultural spectrum between them:

> Once one is committed to the view that science is not so clearly separable from the human sciences . . . or from other humanist enterprises, history of science begins to blur with social history. Literature becomes part of the history of science. Science is reflected in literature. And the tools of literary criticism become instruments in the understanding of scientific discourse.[107]

Nevertheless, hostility to science among the descendants of Snow's "literary intellectuals" can often still be found. Writing in the *Daily Telegraph* in 1991, during one of the more heated periods of debate around the cultural divide, novelist Fay Weldon complained to the scientific community:

> We, the public, have to put up with your brave new world because there is no going back, and the past is an ignorant and brutal land and all of us were miserable there. But don't expect us to like you.[108]

As we shall show in Chapters 5 and 6, science gets a much better deal from the media, arts dominated or not, than might be supposed from some of the more vocal complainants within the scientific community. But if Weldon's views appear to signify that not much has changed, after all, in the content of the exchanges between "literary intellectuals" and scientists, one might look at the change in tone. In the 1880s, Matthew Arnold was brashly confident that science would forever remain secondary to "humane letters"; F. R. Leavis, in the 1960s, pleaded that the human and moral insights of literary culture should not be put to one side in the dash for scientific change and advance; in the 1990s, Fay Weldon was left to lament (from her point of view) that science had got so far ahead. If there is still cultural resistance to science in public, today it may be more a matter of defense than offense. As to the "two cultures," the last word might well be left to Snow himself: he saw it as "more than a dashing metaphor" and "less than a cultural map."

One area of the cultural map that Snow did not explore was the potential divide, and even conflict, between the natural and the social sciences. We look at that in the next chapter.

## Chapter 3

# Popular Science: Friend or Foe?

**Opening science to public scrutiny is a risky business. Popularizing science** may not necessarily make science more popular: if no publicity is bad publicity, then science must be the exception that proves the rule. To know science is not necessarily to love it, and the publics that repeatedly tell survey researchers that they want more information about science are as likely to be working out what to criticize or avoid as what to applaud or embrace.

These days, publics for science include not only "the man and woman in the street" but also groups whose expertise lies in fields such as history, philosophy, sociology, or cultural studies and who bring this expertise to bear on science. While keeping at a distance from the scientific community, they may, like the many amateur astronomers and botanists, develop considerable expertise in the science they study. When the general public reject science, scientists may choose to dismiss the implied criticism as a symptom of ignorance or fear; but when academic "science critics" analyze science, they are using many of the same tools as the scientists, working within the same institutions, and functioning ostensibly according to the same rules of inquiry. Scientists find their criticisms rather harder to ignore.

Some scholarly critiques of science offer scientists much cause for self-satisfaction. Heroic histories and studies of science as culture celebrate the achievements and impact of scientists; and it is surely testament to the power of science that philosophers and sociologists should think it worthy of scrutiny. However, like the public panics over tainted food or unsettled ecosystems, academic studies of science also draw attention to science that is sometimes ill at ease with society and with itself. Some scientists welcome this feedback and use it constructively; others just ignore it. In the early 1990s, some distinguished scientists mounted high-profile campaigns against their critics, thus bringing even more attention to the tensions between science and its publics. This has become, particularly in the United States, an extremely polarized and often bitter debate, and of such intensity that it has become known as the "science wars."

In this chapter we look at these contentious critiques and ask how science might respond. Do we risk too much by opening the doors of science, or is science big enough to take it?

## ANTI-SCIENCE

In the summer of 1988, science was the subject of a publishing phenomenon: for months, a popular science book topped the bestseller lists around the world. British physicist Stephen Hawking's *A Brief History of Time*, which attempted to trace the evolution of the universe back to the very moment of the big bang, sold millions.[1] Hawking himself, who holds the Lucasian Chair of Mathematics at Cambridge University once occupied by Sir Isaac Newton, became a popular icon of modern science. Wheelchair-bound and unable to speak without the help of a voice synthesizer, he epitomized the power of the intellect reaching out from its all-too-frail human frame to encompass the wonders of deep space and time. The message from Hawking was that science—and particularly modern cosmology and high-energy physics—could supply answers to the most troublesome questions of where, in the beginning, we came from and how we came to be what we are, without recourse to theology or mysticism. Hawking's position had not changed nearly a decade later, when he reiterated his views in the 1997 BBC/PBS television series grandly entitled "Stephen Hawking's Universe."

To be sure, Hawking was far from being the first to write in this vein, but the success of *A Brief History of Time* encouraged publishers to sign up other leading physicists to explain their perspective on the big questions. Some of the results were also vehicles for encouraging funding for big science projects, such as Steven Weinberg's *Dreams of a Final Theory* (1993) and Leon Lederman's *The God Particle* (1993). Others, such as Paul Davies's *The Mind of God* (1992), argued that, far from undermining belief, modern science provided the one sure route to religious knowledge and certainty. For the life sciences, Richard Dawkins showed how modern evolutionary theory could explain all the world's biological diversity and even how social behaviors arose, while Daniel Dennett took on the troublesome question of human consciousness. Though none of these authors achieved the sales of *A Brief History of Time*, their books and scores of others amounted to a mini-boom in popular science publishing that (at the time of writing) shows little sign of abating. Coupled with a wide variety of television programs on science and natural history, broadcast typically on the Discovery Channel or on public service channels in the United States and on at least three of the five terrestrial channels in Britain, a rich diet of popular science is on offer to the public.

However, for an important section of the scientific establishment, this apparent success in popularizing the achievements of science is tinged with a bitter taste. For alongside the growing markets for popular science are growing markets for "alternative science" such as homeopathic medicine, crystal healing, and investigations of the paranormal. Legal action by American Christian fundamentalists, who claim that "creation science" deserves the same space in the curriculum as evolutionary theory, has made state educators anxious about teaching children the natural history of the Earth. While astronomers are seeing further than ever, a parallel intensity of interest in the stars among other earthlings seems confined to the astrology columns. Not only that, but science has had to contend with campus criticism from the social and historical sciences. This criticism is, for several champions of the cause of science, of greater concern in delegitimizing science than all the New Age and fundamentalist assaults put together; according to these champions, our scientific and technological society harbors forces which they describe as "anti-science."

The critiques of science lumped together under the umbrella of "anti-science" come from such diverse viewpoints that, were they all to be accepted, science must be both the authoritarian prop of exploitative establishments and the deeply subversive ideology of those gnawing at the vitals of traditional belief and moral society. But do such criticisms add up to an anti-science "movement," coherent or otherwise? And do such criticisms delegitimize science to the extent of discouraging funding for high-energy physics projects or of promoting New Age mysticism or even of inciting physical attacks on scientists who use animals in their research and on doctors working in abortion clinics? Many leading figures in science are very much afraid that they do and have taken up arms—to a greater or lesser degree—against them. The former editor of *Nature* John Maddox, for example, has warned against a "general and benign tolerance of astrology (and other mumbo-jumbo such as faith healing, water divining and spiritualism) apparently on the grounds that they are harmless pursuits."[2]

Among the targets of these scientific opinion leaders are the mass media. The media are accused of scaremongering over questions of nuclear safety or whether our food is fit for human consumption, of not explaining the real nature of risk, and of sensationalizing promising medical developments as cure-alls. For many in the scientific community, the alacrity with which the claims of every spoon-bending mind reader are taken up by, particularly, the tabloid press is acutely irritating, and the daily horoscope is a chronic and weeping sore. In a televised lecture in 1996, Dawkins—speaking in his role as Professor of Public Understanding of Science at the University of Oxford—felt this media preoccupation with "pseudoscience" threatened the proper appreciation of science and dampened "the appetite for wonder" that it could satisfy.[3] Television channels, he warned, were unleashing an "epidemic of paranormal propaganda" that threatened to take us back to "a dark age of superstition and unreason, a world in which every time you lose your keys you suspect poltergeists, demons or alien abduction." High on his list of offenders was the popular TV drama series "The X-Files." Dawkins characterized this program as offering viewers a choice between a rational and a paranormal explanation of the usually violent happenings at the center of the show and invariably abandoning the rational in favor of the supernatural. "Imagine a crime series," Dawkins said, "in which, every week, there is a white suspect and a black suspect. And every week, lo and behold, the black one turns out to have done it. Unpardonable? Of course! You could not defend it."

Dawkins's target on that occasion was just one of a number of strands that, in the opinion of Harvard physics professor and historian of science Gerald Holton, have come together to form a "pseudo-scientific nonsense that manages to pass itself off as 'alternative science'. . . in the service of political ambition."[4] While recognizing that "the term anti-science can lump together too many, quite different things that have in common only that they tend to annoy or threaten those who regard themselves as more enlightened," Holton nonetheless warns that "what the more sophisticated anti-scientists offer is . . . an artificial and functional, and potentially powerful countervision of the world, within which there exists an allegiance to a 'science' very different from conventional science." Moreover, "that countervision has as its historic function nothing less than the delegitimization of science, . . . above all . . . its classic, inherently expansionist ambition to define the meaning and direction of human progress." In Holton's scheme, there are "four most prominent portions of this cohort of delegitimization." These comprise:

(a) "from the intellectually most serious end, there is now a type of modern philosopher who asserts that science can now claim no more than the status of a useful myth," people such as philosopher Mary Hesse and sociologist Bruno Latour;

(b) "a group . . . of alienated intellectuals . . . who in previous centuries would have been among the friends and most useful critics of science" but whom "the fantastic growth rate of new knowledge [has] left impotent," dealing them "a devastating humiliation." As representative of these "powerful intellectuals," Holton has in mind novelist Arthur Koestler and literary critic Lionel Trilling;

(c) "the Dionysians, with their dedication ranging from 'New Age' thinking . . . to crystal power"; and

(d) "a radical wing of the movement . . . [who think] science now has the fatal flaw of androcentrism . . . [and call for] a more radical intellectual, moral, social and political revolution."

It is against this last group that Paul R. Gross, the former director of the Woods Hole Marine Biological Laboratory, and mathematics professor Norman Levitt have directed a concerted broadside. In their combative 1993 book, *Higher Superstition*, they explain:

> Our subject is the peculiarly troubled relationship between the natural sciences and a large and influential segment of the American academic community which, for convenience but with great misgiving, we call the "academic left."[5]

Among this academic left, they claim that there is "open hostility toward the *actual content* of scientific knowledge and toward the assumption . . . that scientific knowledge is reasonably reliable and rests on sound methodology." In adopting this position, the academic left is, according to Gross and Levitt, engaged in rejecting "the strongest heritage of the Enlightenment." They identify five strands that add up to this anti-science position:

(a) postmodernism, which is "grounded in the assumption that the ideological system . . . of Western European civilization is bankrupt and on the point of collapse";

(b) the "traditional Marxist view" that "science is really 'bourgeois' science, a superstructural manifestation of the capitalist order";

(c) the "radical feminist view" that science "is poisoned and corrupted by an ineradicable gender bias";

(d) multiculturalism, which views "science as inherently inaccurate and incomplete by virtue of its failure to incorporate the full range of cultural perspectives"; and

(e) "radical environmentalism," which "condemns science as embodying the instrumentalism, and alienation from direct experience of nature which are the twin sources of an . . . ecological doomsday."

Gross and Levitt deal most sharply with what have come to be known as "social constructivist" theories of scientific knowledge. These theories suggest that scientists, rather than using an infallible method to reveal the true facts of nature, are instead constructing explanatory stories from data produced and interpreted in ways conditioned by, and designed to reinforce, both the scientists' social and cultural mores and their preconceptions and expectations of the natural world—a world that contributes to only a limited extent to the scientists' reconstruction of it in the laboratory. While accepting a "weak" version of such theories—it would be difficult for anyone to deny that science is an activity carried out within human society, with its cultural standards, influences, and preferences—Gross and Levitt reject outright both the idea that science is "but one among many" social discourses and the accompanying claim that "science deludes itself when it asserts a particular privileged position in respect to its ability to 'know' reality." In their polemic against the "academic left," Gross and Levitt accuse their opponents of studying the form of science while being totally ignorant of its content. They argue that while the terms in which science is expressed might demonstrably reflect social forces, sociologists have failed to present any concrete cases of social influence on the eventual outcome—the resulting facts—of scientific debates and researches. In a nutshell, Gross and Levitt say that critics of science fail to show how it would have been otherwise had different social biases (or none at all) been at work.

*Higher Superstition* was a call to arms in the science wars—an injunction on scientists to challenge their obscurantist colleagues in the social and human sciences. According to Gross and Levitt:

> Today we find ourselves confronting an ignorance . . . now conjoined with a startling eagerness to judge and condemn in the scientific realm. A respect for the larger intellectual community of which we are part urges us to speak out against such an absurdity. This, we consider, is one of the duties of the scientific thinker, a duty commonly ignored.[6]

As a result, the public have been treated to head-to-head confrontations between battling dons, giving and expecting no quarter in the struggle to keep their academic standards clear of the mire of ignorance.

In Holton's opinion, a more effective strategy for dealing with the anti-science movement is to examine the *Weltbild*—the personal world-picture—of its adherents and to take advantage of its internal contradictions. Thus, he argues that although each of these world-pictures is

> essentially internally functional in its own terms, [its] constellation of
> underlying beliefs . . . is not necessarily internally coherent or non-
> contradictory. . . . At any given time in a given culture, many clusters
> of partly overlapping individual world-pictures will be discernible.

As a result, "no world-picture is truly anti-scientific": to hold such a position
would be to set one's face against the development of new knowledge altogether.
That, in turn, means that individual world-pictures are "not necessarily stable over
time."

> For both individuals and for a community, changes in allegiances to a
> . . . world-picture can sometimes be seen to be correlated with changes
> in external (e.g. political, economic) conditions that test or challenge
> the functionality of the existing world picture.

So Holton concludes with some optimism that:

> the increase in awareness of internal contradictions, brought about by
> external stress, can provide the opportunity for the most effective edu-
> cating intervention to take place.[7]

Thus, scientists should involve themselves in interactions that "bring to light directly the internal contradictions of the alternative picture," and the education system should provide "curriculum materials specifically tailored to explain the
power and limits of science."

Holton's optimism in this respect is, however, tempered by his own past as a
refugee from Nazi Germany:

> History has shown repeatedly that a disaffection with science . . . can
> turn into a rage that links up with far more sinister movements. Could
> it be, at the end of the century, that the widespread lack of proper un-
> derstanding of science itself might either be a source, or a tell-tale
> sign, of culture's decline?

That world-pictures may not always gravitate toward the scientific at times of
great social change was also outlined by Sergei Kapitza, physicist and academician
of the former Soviet Union. As the power of the Soviet Communist Party weak-
ened, anti-science attitudes grew, in part as a reaction to official insistence that
communism had been based on the theory of "scientific socialism" and in response
to technological disasters, of which Chernobyl was merely the most explosive.
Writing in 1991, just prior to the break-up of the Soviet Union, Kapitza reported:

> . . . there is a strong correlation of anti-science and anti-technology
> trends with publications on sex, violence and extreme social ideas,

such as rampant nationalism and fascism . . . the spread of these ideas
is not at all harmless . . . powerful irrational forces are at work, perhaps
supported by obscure political motives.[8]

For Kapitza, the period of the 1990s in Russia recalls Europe of the Reformation
and the "Thirty-Year War": "during this upheaval, superstitions of all kinds flourished . . . reportedly 50,000 witches were burned alive or drowned" in the name
of defending the Christian faith.

    These are just a few examples of the kinds of terms in which distinguished
scholars have expressed their concern over what they see as the anti-science *Zeitgeist* pervading Western culture. But are they right to be so worried? At the heart
of the issue is that the critics and "delegitimizers" have raised in peoples' minds
the issue of trust: How far can the public trust science to provide reliable, objective, and unbiased knowledge? How certain is this knowledge—will it stand the
test of time? And how deep is this knowledge—does science have its limits or can
it give the ultimate answer? In what follows we look at some of the more considered and persuasive critiques of science and ask whether science can survive this
perceived anti-science onslaught, on campus and in public.

## THE SOCIAL CONSTRUCTION OF SCIENCE

The power and limits of science are matters of considerable debate. Defining
them is just part of the problem; whether or not to communicate them is just as
large a concern. For while the scientific community enjoys its reputation as a
trustworthy solver of problems, it also berates both the public for unrealistic expectations and exaggerated fears and the media for fostering extreme views of
what science can and cannot do. Just how accurate a picture the public gets of science and scientific processes, and of the people who do science, has concerned
many authors. Television popularizations of the life and work of great scientists
such as Louis Pasteur and Charles Darwin often depict them as men struggling to
make sense of a mass of information but impelled, nonetheless, inexorably toward
their ultimate triumph over nature: there is a defining "eureka" moment in which
they see the light. A brief scene or two depicts the enunciation of their great theory, and then their new idea sweeps all before it; disciples multiply; and, at the end
of their lives, the Great Men of Science look back at the new avenues that they
have opened for research, discovery, and human progress. Popular accounts of the
rise of Einstein's relativity theory begin with the seeming inviolability of New-

ton's grand mechanical scheme of the universe; but gradually it becomes clear that the clockwork universe is failing crucial experimental tests. Einstein brings his razor-sharp mind to the rescue of physics. Working logically from the very fundamentals of science, needing only a pen and the backs of menus to forge his new vision, the humble Swiss patent clerk succeeds in effecting a revolution in scientific thinking.

Sociologists of science see a rather different scene. In *The Golem*, Harry Collins and Trevor Pinch depict science neither as a careful distillation of observation and experiment nor as the working out of a profoundly logical process. And, for them, science is certainly not unambiguously the obedient servant of rational humanity, the provider of deep insights and helpful technology. Collins and Pinch's science is "a lumbering fool who knows neither its own strength nor the extent of its clumsiness and ignorance."[9] The golem of science, taking its name from a creature of Jewish mythology, is a powerful beast. "It will follow your orders . . . . But it is clumsy and dangerous. Without control a golem may destroy its master with its flailing vigour." The golem was a creature of clay that was animated by having the word *emeth*, Hebrew for truth, inscribed on its forehead. "But that does not mean it understands the truth—far from it," they explain.

Collins and Pinch describe a series of episodes in science in which the path by which controversies are resolved or new theories validated are a million miles, they claim, from the clean-cut methods and logical processes of science that are usually portrayed to the public. In one example, the worlds of psychology and biochemistry were turned upside down by claims that memories could be passed by feeding one animal to another, or by injecting homogenized tissue from one animal into another—claims made first for simple planarian worms in the 1950s and then for rats and mice in the 1960s. The prospect was raised of aspiring actors cannibalizing the stars of stage and screen, or students their prize-winning professors, instead of learning their lines or their notes. Research biologists in the field were divided between those who claimed they could train worms and then pass on this training simply by feeding mashed worm to other worms and those who could not. The latter group called the former "fraudsters"; to the former, the latter were simply "incompetent." In this instance, the controversy was eventually resolved, say Collins and Pinch, because the respective scientific fields simply moved on. One of the champions of "edible memory" died, and another's work was eventually considered so irrelevant by his peers that his laboratory closed for lack of resources. But there was neither a "defining moment" nor a "breakthrough" theory or experiment that clinched the matter. Edible memory went out with a barely audible whimper.

Collins and Pinch have some equally interesting and provocative things to say about the question of "cold fusion," the claim by chemists Stanley Pons and Martin Fleischmann that nuclear energy could be released by a simple electrochemical reaction in a test tube. When this discovery was announced at a press conference in March 1989, it generated huge headlines and research funds around the world. If they were right, Pons and Fleischmann had just delivered to the world electricity that was genuinely "too cheap to meter." The commercial potential appeared to be enormous, and the lawyers got busy. But soon it became clear that Pons and Fleischmann were reluctant to publish their results in a refereed journal, which would open their work to scientific scrutiny. After many renowned laboratories failed to reproduce cold fusion, the scientific community at large came to the conclusion that this was a case of "bad science" in which the unusual use of a press conference had been allowed—unforgivably—to replace the normal peer-review process. However, Collins and Pinch see it somewhat differently:[10]

> Pons and Fleischmann appear to have been no more greedy or publicity seeking than any other prudent scientists would be who think they have on their hands a major discovery with a massive commercial payoff. The securing of patents and the fanfare are inescapable parts of modern science.

This matters for the public because it is only in exceptional circumstances like those surrounding the cold-fusion episode that the public find out how science really works; only then do people get to know "what everyone should know about science," to quote *The Golem*'s subtitle. And such knowledge is generally disturbing:

> The shock comes because the [public's] idea of science is so enmeshed in philosophical analysis, in myths, in theories, in hagiography, in superstition, in fear, and most important, in perfect hindsight, that what actually happens has never actually been told outside of a small circle.

According to Collins and Pinch, if the scientific community, in cahoots with the media, thought it was doing the public a favor by not troubling them with the complexities of scientific research as it is really carried out, it was mistaken. The shock of being disabused of the simple picture of science is a vital one. "The point is that for citizens who want to take part in the democratic process of technological society, all the science they need to know about is controversial": so it is the mess, the disagreements, and the uncertainties of science that matter most in the public sphere. After all, when science is running smoothly, we can all afford to ignore it.

If Collins and Pinch challenge the simple picture of how new scientific knowledge is gained, Bruno Latour takes on public perceptions of the very ethos of scientific research. Latour challenges the idealistic vision of science set out in the 1940s by the sociologist of science Robert Merton, who defined the norms of scientific research as being "communism"—the fruits of science are collectively the property of all scientists; "universalism"—what is true for one scientist is true for all, irrespective of race, religion, personality, or culture; "disinterestedness"— the search for truth is conducted without fear, favor, or preconception; and "organized scepticism"—a constructive distrust of new results is encouraged, which promotes critical testing and evaluation.[11] Thus, Merton's ethic of science was CUDOS, an ideal with implications, and one often put forward to the public as the basis for "good science." However, in his *Science in Action*, Latour depicts science as the doubled-headed god Janus, saying one thing retrospectively and another when facing the future—one thing to the insiders of the scientific community and another to the waiting public.[12] Scientists, in Latour's approach, have to combine the qualities of generals and diplomats, marshaling their arguments, results, and equipment like troops on the battlefield, while tying up in alliances as many of their peers as possible. This Machiavellian image of science as war of all against all is clearly a long way from the disinterested, gentlemanly collaboration envisaged by Merton.

In introducing his study, Latour speaks of opening and closing "Pandora's Black Box" of scientific research and controversy:

> We will enter the facts and machines while they are in the making; we will carry with us no preconceptions of what constitutes knowledge; we will watch the closure of the black boxes and be careful to distinguish between two contradictory explanations of this closure, one uttered when it is finished, the other uttered while it is being attempted.[13]

Latour's aim, therefore, is to look at "science-in-the-making" by following the processes that lead up to the scientists involved deciding on "the facts of the matter" and agreeing to close the debate, to shut the lid on the "black box."

One of the controversies Latour describes is the dispute over the nature of the hormones released by the brain's pituitary gland—growth hormone-releasing hormone (GHRH), championed by Shally, and the rival growth-related factor (GRF) of Guillemin. The "winner" of the dispute has, as a potential prize, a discovery that would help treat growth illnesses such as dwarfism. During the course of this dispute, Guillemin manages to show that Shally's GHRH has an amino acid structure identical to that of hemoglobin (a constituent of red blood cells) and

that Shally has therefore allowed himself to be misled by contaminated specimens. Latour points out that although Guillemin has struck negatively at his rival,
he has not positively enhanced his own position. As the dispute wends on, both
sides launch attack and counter-attack. These involve carrying out more and more
detailed analysis and experimentation. As such, they require setting up more and
more complicated and better equipped laboratories, until one side can no longer
afford to keep up "marshaling troops for the battle." The publication strategies of
the rival laboratories involve citing the papers of more and more of their fellow
workers in support of their claims, while getting these same fellow workers to cite
*their* papers favorably, thus making allies and at the same time marginalizing (and
eventually putting out of business) those who do not fall into line.

During the course of his analysis, Latour draws out a "rule of method" (one
of many he puts forward) for analyzing the reality of scientific research, which is
of prime significance for his whole approach:

> Since the settlement of a controversy is the cause of Nature's represen-
> tation not the consequence, we can never use the outcome—Nature—
> to explain how and why a controversy has been settled.[14]

To emphasize the point, Latour's backward-looking Janus face—the one uttering
the reason for closing a controversy box after settlement—says: "Nature is the
cause that allowed controversies to be settled." This is the face that is usually presented to the public; scientists eventually get to the truth of Nature. But the for-
ward-looking face, which represents what is being said during attempts to close a
controversy box—and hence is rarely on public display—says: "Nature will be the
consequence of the settlement."

Latour's argument is that if we wish to understand how scientific controversies are resolved, we should not expect to see them ending when scientists eventually reach the truth, for this "truth" is itself the construct of the activity of the
community of scientists. In this "social constructivist" approach, what counts are
the very factors that Merton's CUDOS would lead one to discount—private control of knowledge and, particularly, of the equipment needed to make knowledge;
truth for one laboratory being error for another; self-interestedness in the outcome
of the dispute, with the kudos and resources that will likely accrue to the winner;
and appeals to fellow scientists to make their judgments as to whom to believe on
the basis of the power and reputation of the groups putting forward the contending
theories. The importance of this for the public is clear where, for example, the science that is being undertaken involves deciding on whether or not something is a
health hazard, where that science is both new and controversial, and where there

are vested interests involved in the outcome. We look in more detail at such cases in Chapter 7.

Both *Science in Action* and *The Golem* critique the public image of science from the standpoint that real science owes a lot (if not all) of its character to the social practice of scientists; by and large, they concentrate on internal (or micro) social pressure (although external resourcing is clearly a factor). Both books are broadly in the tradition of what has come to be known as the "strong program" of the sociology of scientific knowledge. One of the key centers of critiques of science from within the strong program has been the University of Edinburgh, from where David Bloor published *Knowledge and Social Imagery* in 1976. Bloor's book attempts to analyze the (macro) social influences essentially external to science that shape its development. He raises questions as to whether scientists as scientists inevitably absorb the spirit, common sense, and culture of their times—and, if so, how much this affects their work and its framing.[15] Examples might include the preference of English geologists for gradual uniformitarianism in the early 19th century as against the abrupt catastrophism of their revolutionary neighbors across the Channel and the role that the skepticism and disillusionment of the 1920s may have played in giving rise to the "uncertainty" and probabilistic nature of quantum mechanics.

Traditional histories of science have depicted the evolution of scientific ideas as a gradual, or sometimes cataclysmic, replacement of incorrect theories by ones that are closer to the truth. Thus, Stahl's phlogiston theory of burning was replaced by Lavoisier's understanding that combustion involves combination with a component of air called oxygen. This idea of "oxidation" is now interpreted more generally by chemists in terms of the transfer of electrons between atoms. Error has been replaced by truth, and truth by more general truth. Bloor, however, considers this an insufficient explanation for the history of science and demands "symmetry":

> The strong programme enjoins sociologists to disregard [truth] in the sense of treating both true and false explanations alike for the purposes of explanation.[16]

For the purposes of sociological research, then, this puts beliefs in witchcraft and how magic powers are inherited on the same footing as a scientific understanding of genetics or mathematics. Social processes are at work in the formation of all beliefs, and it is the job of the sociology of knowledge to examine the sociological aspects of those processes, not to judge whether or not the particular beliefs are "true." The symmetry principle allows sociologists, for the purposes of their inquiry, to suspend disbelief and to treat all forms of knowledge as valid within the circumstances in which they arose. This is not to say that the sociologists themselves believe that all forms of knowledge are equally valid: the symmetry principle is a methodological tool, not a judgment on the relative value of ideas. Academically at least, the sociologists leave the value judgments to others.

For Bloor, while the material world produces "causal promptings" that appear in experimental results, and thus in scientific ideas, theories are, at heart, "social conventions" accepted by scientists:

> The question may be pressed: does the acceptance of a theory by a social group make it true? The only answer that can be given is that it does not. There is nothing in the concept of truth that allows for belief making an idea true. Its relation to the basic materialist picture of an independent world precludes this. But if the question is rephrased and becomes: does the acceptance of a theory make it the knowledge of a group, or does it make the basis for their understanding and their adaptation to the world?—the answer can only be positive.[17]

Thus, there may be a gap between truth and these social conventions, although Bloor is keen to emphasize that recognizing that there are social influences shaping scientific knowledge does not necessarily make that knowledge wrong or invalid. The "role of ideas" and "social factors" are not in competition.

Bloor accuses those who would deny social science the right to investigate knowledge, and how it is formed, of mystification. And he goes further, saying that there is a tendency to mystify knowledge when the guardians of that knowledge feel under a social threat. He contrasts the 18th century social commentator Edmund Burke's "mystification" of society and politics under the perceived threat from revolutionary France with the naturalistic approach of the Enlightenment that preceded 1789. Similarly, he contrasts philosopher Karl Popper's falsification theory of scientific discovery, which takes place almost entirely in the realm of logic, with the naturalistic—and more sociological—approach of historian Thomas Kuhn, involving "paradigms" that are periodically "shifted."

In *The Logic of Scientific Discovery*, Popper advanced the idea that the only road to scientific knowledge was through deducing it from the principles set out in theories.[18] These theories—Popper did not feel obliged to detail how they arose—could only be falsified, never proved true. Induction—building up theories from observing a large number of instances of the same effects or behaviors—was unreliable as a method of deriving knowledge, said Popper, since one could never know when one would come across an instance of the effect or behavior not being observed, thus falsifying the theory. Kuhn's more sociological proposal was that scientists by and large worked within disciplinary paradigms, or models, during their day-to-day activities. Every so often, however, the internal contradictions of the model, or a set of particularly compelling observations, caused a revolution, a "paradigm shift" that resulted in the formulation of a new model.[19] The change from Newtonian mechanics to relativity is often cited as just such a paradigm shift. Popper, says Bloor, formulated his cautious and rather pessimistic ideas during the uneasy 1930s and 1940s, whereas Kuhn's bolder interpretation of the history of science was a product of the more secure post-war years.[20]

From his discussion of theories of the history and development of science, Bloor goes on to discuss the way in which society shapes science itself. The medieval cosmological model had the Earth at the center of a series of concentric spheres:

> In a . . . sense it was a social phenomenon, an institutionalized belief, a part of culture. It was the received and transmitted worldview, sanctioned by the authorities, sustained by theology and morality and returning the service by underpinning them.[21]

To demonstrate the deeper social influence on science, Bloor discusses the Greek conception of mathematics. As an example, he claims that Greek dialectical logic, classifying numbers as either odd or even, led them to deny that 1 was a number. Along with other examples, this shows that even the Greeks—much vaunted as founders of modern Western science—had an "alternative" mathematics.

Some of the most pungent academic critiques of science owe their derivation to the philosophy of postmodernism.[22] It is these critiques that cause the deepest concern among scientists and are most heavily criticized by people like Gross and Levitt, since they go to the heart of the question as to whether or not science is reliable and whether or not the public should trust it. One of the earlier strands of postmodernism was the criticism of modern architecture, taking on its grand schemes which—the critics claimed—failed to take account of the human dimension and the diversity of the people expected to live or work in its buildings. However, postmodernism quickly turned its attention to other spheres, taking issue with the modern idea of "progress" handed down from the Enlightenment. Grand schemes were to be swept away in the name of diversity and "relativity"—why should any one point of view be preferred above another? In the realm of politics, one of the chief targets was Marxism—at least in the Stalinist version that prevailed in the Soviet Union and its satellites. But science, too, was a natural target. Science was motivated by a "grand scheme" which promised progress; science replaced ignorance with progressively more detailed knowledge; and, most importantly, science progressively replaced error with truth. In the postmodern world, it would therefore appear, belief in the mythology of dragons and demons is just as useful socially in some cultures as adherence to the science of volcanology and field theory is in others—indeed, some postmodernists claim that science is just the myth of modernity.

One of the postmodernists who has discussed science in detail is Jean-François Lyotard, who summed up his position as "incredulity towards meta-narratives."[23] Although science itself is not cited as one of Lyotard's grand or meta-narratives, it relies for its legitimacy, he claims, on the Enlightment's grand narratives of "truth" and "progress." But these narratives are precisely what the postmodernists claim to have demolished, and thus: "We no longer have recourse to the grand narratives—we can resort neither to the dialectic of Spirit nor even to the emancipation of humanity for a validation for postmodern scientific discourse." As a result, "science, far from successfully obscuring the problem of its legitimacy, cannot avoid raising it with all of its implications, which are no less sociopolitical than epistemological." And it is impossible for scientists even to negotiate among themselves a consensus on the truth, since "consensus is a horizon that is never reached."[24]

## SCIENCE AS POWER

If it is once admitted that science and scientific knowledge are affected by social influences, it is then possible to investigate the role that different factors play in the process. Thus numerous works have looked at how the way in which society is structured—the distinctions of class, gender, and race, the divisions between rich and poor nations—make themselves apparent in scientific ideas and in the role of science within society. One of the most ambitious of these was the now largely ignored *Science in History* (1954), by the crystallographer and Marxist J. D. Bernal. Marx and his close companion Friedrich Engels were themselves very enthusiastic about science and had drawn out from the latest scientific discoveries and theories elaborations of their dialectical materialist philosophy and support for their scientific approach to history and society.[25] While Bernal's book made close links between prevailing scientific ideas and technological developments and the social structures into which science fitted, it continued the "Marxist" tradition of a generally positive critique of science.[26]

The arrival of postmodernism, however, with its questioning of the "grand schemas" of both Marxism and science, means that many writers who consider themselves to be following in the left-socialist tradition have adopted a far more negative attitude toward scientific ideas and the social functions of science in the late-20th-century, free-market economy. In his *Science as Power*, Stanley Aronowitz traces the evolution of Marxist and post-Marxist attitudes to science. Aronowitz's book is itself a critique of the way in which "scientific practices promote a universe in which the domination of nature is linked to the domination of humans." He talks of "the imputation to nature of characteristics that are nothing more than the objectivization of the table of organization of the social world." As well as being used to justify power relations, science has its own power. According to Aronowitz:

> The power of science consists, in the first place, in its conflation of knowledge and truth. Devising a method of proving the validity of propositions about objects taken as external to the knower has become identical with what we mean by truth.[27]

To consolidate its independent power, science has sought to separate itself from the theory of science, insisting that history, philosophy, and sociology of science have no contribution to make to the development of "pure" science.

For the past century, according to Aronowitz, it has been apparent that science is "central for the processes of economic reproduction" in terms of its importance for developing new technologies. At present, science is subordinate to the state in capitalist economies, but this situation is changing as a result of the activity of the "knowledge communities." "Science is a language of power," Aronowitz argues, "and those who bear its legitimate claims, i.e. those who are involved in the ownership and control of its processes and results, have become a distinctive social category equipped with a distinctive ideology and position in the post-war world." That scientific knowledge communities regularly declare themselves to be neutral on political questions shows "a studied naïveté concerning the implications of accepting resources made available by the state for research." Aronowitz calls science "the discourse of the late capitalist state." "Science is rapidly displacing, as dominant discourse, the old ideologies of the liberal state . . . . This development coincides with the consolidation of bureaucratic power, . . . the ostensible social form of scientific rationality." To avoid this bleak future of an overpowering scientific bureaucracy, "an alternative science would have to imagine, as a condition of its emergence, an alternative rationality which would not be based on domination."[28]

The feminist movement has been among the most active in attempting to create the kind of "alternative rationality" called for by Aronowitz. Feminists have sought to "rediscover" the role women have played in scientific research and development—whether as individuals or as a group—and the obstacles, in the form of educational and scientific structures, that are placed in their way. The scarcity of women in science, and their subordinate position in Western society, results in "androcentrism" and sexist bias, both in the nature of the scientific enterprise and in the way that nature is portrayed and investigated. The most thoroughgoing feminist critiques attempt to expose the androcentrism of scientific epistemology and to inquire whether it is possible to produce a theory of knowledge for science either from a feminist standpoint or from one that is entirely gender-free. Explaining these approaches, sociologist Sandra Harding points out that science is

> not as monolithic or determinist as many think. . . . It has been versatile and flexible enough throughout its history to permit constant reinterpretation of what should count as legitimate objects and processes of scientific research . . . .[29]

Harding's own analysis of feminist critiques of science divides them into those that attack "bad science" and those whose target is "science as usual." In anthropology, for example, "basing generalizations about humans only on data about men violates obvious rules of method and theory" and is the sort of "bad science" that deserves criticism. Feminist critics who follow this approach generally believe that the "scientific method is supposed to be powerful enough to eliminate any social biases that might find their way from the social situation of the scientist into hypotheses, concepts, research design, evidence gathering, or the interpretation of the results of research" if only the method is applied rigorously.

The criticism of "bad science" flows from the epistemology that Harding calls "feminist empiricism." This approach "is often thought to be less threatening to the practices of the sciences." "It is indeed conservative in several respects, and these constitute in part both its strength and its weaknesses." For Harding, the main shortcoming of the feminist empiricists is that

> on the one hand, they try to respect the dogma that one can explain "good science" without referring to its social causes. On the other hand, they think the women's movement is a cause of "better science."[30]

Harding herself inclines toward the critique of "science as usual" and the accompanying theory of knowledge that she calls "feminist standpoint epistemology."

> Knowledge is grounded in particular historical social situations . . . . In societies where power is organized hierarchically—for example, by class, race or gender—there is no possibility of an Archimedian per-

> spective, one that is disinterested, impartial, value-free, or detached
> from the particular, historical social relations in which everyone par-
> ticipates.

Thus, in feminist standpoint epistemology,

> the distinctive features of women's situation in a gender-stratified so-
> ciety are being used as resources in the new feminist research. It is
> these distinctive resources, which are not used by conventional re-
> searchers, that enable feminism to produce empirically more accurate
> descriptions and theoretically richer explanations than does conven-
> tional research.[31]

One particular aspect of this standpoint, Harding says, is the claim that women's perspectives "mediate the divisions and separations in contemporary Western cultures between nature and culture."

The products and application of scientific research in the form of technology have also been the subject of much debate from the standpoint of feminism. "Throughout these debates there has been a tension between the view that technology would liberate women—from unwanted pregnancy, from housework and from routine paid work—and the obverse view that most new technologies are destructive and oppressive to women," explains Judy Wajcman in her *Feminism Confronts Technology*. Wajcman feels: "The sociology of technology can only be strengthened by a feminist critique. This means looking at how the production and use of technology are shaped by male power and interests." Reflecting the "tension" of the debate, however, she proposes:

> The search for a general feminist theory of technology, or of science,
> is misguided. Instead, I have argued that we need to analyse the spe-
> cific social interests that structure the knowledge and practice of par-
> ticular kinds of technology.[32]

This tension is apparent throughout Wajcman's analysis. Among the topics she discusses, the role of technology in housework and in contraception, pregnancy, and labor features large. Technology began to play an increasing part in childbirth when, during the 18th century, the use of forceps during difficult births became more common. This meant that women were "delivered into men's hands," since only members of the male medical profession were allowed to use surgical instruments. Traditional female midwives were increasingly relegated to a secondary position. Other technological developments in childbirth have increased this dependency of women on the medical profession, reducing them—some claim—"to the status of reproductive objects." But Wajcman points out that many of the technical advances—such as the use of gas during labor—have in fact been demanded by women to free themselves from unnecessary pain and suffering. She criticizes many feminist critiques for treating women as merely passive recipients of whatever technology was available, rather than seeing the active role they played in shaping the development and implementation of those technologies. Similarly, while feminists have rightly highlighted the concentration of research on female rather than male contraceptive methods, sometimes with less than due regard for women's overall health, Wajcman points out the liberating effect of techniques such as the contraceptive pill. Her analysis of domestic appliances is similarly double-edged. "Identifying the gendered character of technology need not lead to a rejection of existing 'patriarchal' technology. Neither does it require us to abstain from working 'in' technology," she concludes.[33]

Mirroring these feminist critiques of science is another growing body of literature that criticizes the white Eurocentrism of science (insofar as most American scientists are also of European, rather than indigenous, African, or Asian, origin). Some of these concentrate on the overimportance traditional histories of science give to the European scientific revolution and Enlightenment and find support in such classic works as those of Joseph Needham on science in China. There are also critiques of the way the role of nonwhite scientists has been belittled; for example, they ask whether the ethnic origins of the great Indian physicist Chandrasekhar diminished his standing in the 1930s debate with English astronomer Sir Arthur Eddington on the evolution of stars, even though he was later proved to have been correct. The difficulty of black scientists in gaining acceptance, compounded if they are also women, has been highlighted by the physicist-turned-historian of science Evelynn Hammonds. "It was the isolation," she says. "It was the fact that Black scientists are questioned more severely. Our work is held up to greater scrutiny; we have a difficult time getting research, getting university positions."[34] Black Americans are, nonetheless, being encouraged to enter science, if sometimes for rather negative reasons. "Understanding the political and economic aspects of science and technology clarifies the need for Black people to be in a position to thwart undesirable innovations and consequences arising from these forces," says Robert C. Johnson. He calls for

> national black organizations . . . to identify and pull together groups of the committed, concerned and capable Black scientists, technologists, engineers, policy analysts, social scientists, community workers, and activists, to review these situations and to devise plans, strategies, and tactics for coming to terms with them.[35]

Taking the anti-Eurocentric critique one stage further, the export of science and technology to developing countries—C. P. Snow's best intentions (see Chapter 2) notwithstanding—has been more recently viewed as yet another form of Western imperialism, a means of keeping former colonies under the economic yoke of Europe and America. This has been effected, critics say, in a variety of interlocking ways. First, the export of technology to countries without an existing scientific and technological infrastructure has placed emerging nations in a financial and knowledge debt to the West. Second, the inappropriateness of much that was exported has resulted in the disruption of traditional means by which the population gained its livelihood: replacing food crops by cash crops, for example. Third, much of the exported technology has—whether by design or not—reinforced the developing world's role as a producer of primary goods, subject to violent (and, more often than not, downward) movements in price. Fourth, much of the exported technology has been military hardware, which has destabilized regions where international and intertribal relations were already tense. This has also made the developing world an arena for the "big powers" to play politics. One interesting take on the civil war in Nigeria in the 1960s, for example, has the U.S. State Department backing Biafra in an attempt to break Europe's stranglehold on Nigerian oil and its support for the Lagos government. Fifth, in this by no means exhaustive list, the intellectual and ideological components of science and technology exports have had the effect of devaluing local knowledge and ways of understanding, often to the detriment of those left to carry on once the Western "experts" have pulled out.

In his passionate description of the problems of famine-torn Africa in the 1980s, journalist Lloyd Timberlake says: "Africa is dying because in its ill-planned, ill-advised attempt to 'modernize' itself it has cut itself in pieces."[36] Some of his most trenchant criticism is reserved for the big dam projects that were very fashionable during the 1960s and 1970s. These, among other things, plunged many African and Asian nations into huge debt—so much so that even the World Bank, which financed many of these schemes, now officially opposes most of them. Nor did the dams, in too many cases, bring the increased production and wealth that was promised. As an example, Timberlake cites projects from the 1980s to provide irrigation in the basin of the River Senegal by constructing two dams that not only displaced local people and gave them in return a meagre seven percent of the resulting irrigated land, but also destroyed woodland and traditional grazing areas. Timberlake concludes with an appeal for appropriate irrigation schemes, based on more locally appropriate technology:

> Despite all the problems with big dams, Africa badly needs hundreds of thousands of new dams and water control projects to get water to the right place at the right time. But the dams required are small ones—not because small is beautiful, but because small is manageable. Systems can be changed and rebuilt as change and repair become necessary, by groups of farmers rather than outside contractors and expensive machinery.[37]

Criticisms of Western science and technology overlap with environmental and ecological concerns, which have traditions stretching back well into the last century but have come particularly to the fore since the 1960s. Hailed as a pioneer of environmentalism, Rachel Carson wrote of a formerly idyllic town in the heart of America:

> . . . the roadsides, once so attractive, were now lined with brown and withered vegetation as if swept by fire. These, too, were silent, deserted of all living things. Even the streams were now lifeless. No anglers visited them, for all the fish had died. No witchcraft, no enemy action had silenced the rebirth of new life in this stricken world. The people had done it themselves.[38]

Carson's book *Silent Spring* made an enormous impression on people throughout the world when it was published in 1962. From then on, a global, if disparate, environmental movement developed, its growth accelerating rapidly in the 1970s. There is an enormous literature in this area, but the fundamental problems, according to the environmentalists, are overpopulation and, even more central, continuous economic growth based on science and technology.

Concerns about the predominance of the "domination of nature" in scientific attitudes are at least part of the raison d'être of the environmental movement. Nevertheless, the more organized sections of it have recently been more willing to counter government and corporate claims about technology with their own scientific evidence. As we discuss at greater length in Chapter 6, in 1995 Greenpeace managed to scuttle plans by the Shell oil company to sink the obsolete Brent Spar oil rig off the coast of Scotland by carrying out their own tests (which, in the event, turned out to be faulty) to show that the rig contained high levels of potential contaminants. Public anger at Shell's apparently cavalier attitude forced a humiliating retreat by the company. However, despite Greenpeace's success in using its own science against Shell and other multinational corporations, some sections of the environmental movement remain deeply hostile to science as it is currently practiced. In cahoots with global capital, science and technology are portrayed as raping the planet and threatening to violate unspoiled wildernesses and to destroy the delicate balance of the global ecosphere on which we depend for our existence. That said, alternative technologies, based on the sensitive application of science in the appropriate social and environmental context, are often invoked as a way out of the present ecological mess.

## SCIENCE AND THE SPIRIT

Criticisms of science as part of the established power system, responsible for much social and environmental damage, have come mainly from radical, and sometimes left-leaning, standpoints. However, the scientific enterprise has been often upbraided from a right-wing, establishment position for its radicalism and for its potential, real and imagined, to destabilize society. Nowhere has this been more the case than in debates over the relationship between science and religion. Ever since Giordano Bruno was burned at the stake for supporting the Copernican view of the solar system and Galileo was shown the Inquisition's torture instruments for popularizing it, science and religion have had a love–hate relationship into which a liberal helping of real or pretended indifference has been mixed. Armed with their particularly powerful way of knowing about the world, scientists have often wondered if they have the means of answering some of the really big questions—"how did we get here, and, more importantly, why?"—that have traditionally been the province of other belief systems, particularly religion. When scientists have tried to answer such questions, they have either been applauded by religious authorities—as they were in the early 19th century when leading naturalists and geologists sought to demonstrate that the newly emerging sciences were rooted firmly in natural theology—or they have been denounced and accused of overstepping the bounds of what are proper subjects for scientific investigation.

Scientists persistently evoke the name of God to assist the popularization of their work or their subject. Thus, Stephen Hawking famously used a holy metaphor at the end of *A Brief History of Time*—despite the fact that he spent the rest of the book demonstrating that God is, at best, irrelevant to modern physics—to express the hope that high-energy physics would lead to an ultimate and unifying theory of the universe:

> Then we shall all, philosophers, scientists, and just ordinary people, be able to take part in the discussion of why it is that we and the universe exist. If we find the answer to that, it would be the ultimate triumph of human reason—for then we should know the mind of God.[39]

This last sentence alone ensured *A Brief History of Time* considerable notoriety. In 1992, Berkeley scientist George Smoot went one better. He was part of the team that announced that a previously little-known NASA satellite called CoBE, the Cosmic Background Explorer, had picked up ripples from space dating from the time of the big bang. The satellite's images were awe-inspiring, and Smoot ensured that his name would be most prominent on the front pages. "It was like seeing the face of God," he said, and journalists dutifully scribbled.

The CoBE results, which are discussed in more detail in Chapter 6, landed coincidentally in the middle of considerable public discussion in Britain about science and its relation to religion and belief. The theme had been aired frequently in the science columns of the broadsheet newspapers and was the subject of a debate at the Edinburgh Science Festival between evolutionary biologist Richard Dawkins and former biochemist John Habgood, the Anglican Archbishop of York. Dawkins divided religious believers into three categories: the "know-nothings," the "know-alls," and the "no-contests." "Faith," he said, "is the great cop-out, the excuse to evade the need to think and evaluate the evidence." Developing the tradition of T. H. Huxley, he continued: "The more you understand the significance of evolution, the more you are pushed away from the agnostic position and towards atheism."[40] Habgood's response was that science was essentially irrelevant to belief:

> . . . whether there was a Big Bang or whether there will be a big crunch makes no difference to theology . . . . Religious experience points to truths that elude [a scientific] kind of treatment.[41]

Habgood's point of view is broadly supported by theoretical physicist-turned-Anglican priest John Polkinghorne. He likens discussions of the meaning of life from purely scientific standpoints to "the ancient tale of human presumption in building the Tower of Babel." "It is time to explore the alternative answer and consider the possibility that it is to God alone that the ultimate cause of life belongs," he concludes.[42] Polkinghorne and Habgood notwithstanding, many scientists have not been put off from investigating the "ultimate cause of life" from a scientific point of view. Anthropic cosmological principles have been proposed and—in their strong form—used to demonstrate that the universe is set up the way it is so that humans can exist and get to know about it. Adding to this, cosmologist Paul Davies believes that our ability to use scientific methods to study the cosmos, and, in particular, the efficacy of mathematics for doing this, leads inevitably to the conclusion that mathematically informed reasoning went into the creation of the universe. In 1995, Davies was awarded the $1 million Templeton

Prize for services to religion; in his personal gospel, cosmology leads to a physical concept of God, while quantum mechanics and chaos theory give modern scientific explanations to the Christian concepts of an indeterminate world in which humans have free will.

Much of this debate has been conducted in very public arenas—at prestige conferences, in the press, and on television and radio. So what is the nonscientific public to make of scientists' claims that they can discuss the deep questions of existence in a meaningful way? During the 1992 crescendo in this discussion in Britain, some people found it all too much to bear. One such was Bryan Appleyard, then a pundit on *The Times*, who gave voice to his exasperation at the Hawkings, Dawkinses, and Davieses of popular science:

> Scientists . . . are at last coming clean: they, and only they, hold the key to the meaning, purpose and justification of human life. The truth or otherwise of this claim, as well as the question of whether it is the real underlying belief by which we conduct our lives, is the most urgent issue of our age. Indeed, I believe it is the only issue of our age, the decisive debate which shapes all others. If we do not begin to understand science, we cannot claim to understand the present.[43]

Appleyard expanded his views later that year in a book entitled *Understanding the Present*. This may not have been a model of logical and rational debate, nor did its critics have much praise for its historical accuracy. But its stridently presented sentiments chimed with those of many in the nonscientific intellectual and literary worlds, who were beginning to feel alarmed or marginalized by the successes of their scientist peers in capturing the popular imagination. In one sense, Appleyard simply said out loud (and rather crudely) what others were whispering. The main thrust of his argument was that it is the basic nature of scientific inquiry and what it says—and does not say—about the knowledge it produces that is responsible for moral and social decline, rather than the debilitating or polluting effects of its sibling technology (though you certainly could not have one without the other). At the center of Appleyard's argument is the following assertion:

> Science is not a neutral or innocent commodity which can be employed as a convenience . . . . Rather it is spiritually corrosive, burning away at ancient authorities and traditions. It cannot really co-exist with anything.[44]

This is not due to merely counterposing one explanation of the world against another, but is the result of the relative nature of scientific knowledge:

> At any one time scientific man can only regard his knowledge as pro-
> visional because something more effective might come along.

This leads to an "enforced neutrality of scientific liberalism" which "makes it progressively more difficult to sustain either morality or spiritual conviction." "We
agree to differ" becomes the order of the day, and "because of the aridity of such
conclusions, even the energy to differ expires" and "tolerance becomes apathy."
The resulting "modern Western . . . liberal society," with all its ills, "is a realization of the scientific method." We are now, Appleyard says:

> . . . in possession of an unprecedentedly effective way of understand-
> ing and acting called science [which] supplants religion and culture yet
> does not answer the needs once answered by religion and culture. . . .
> Science is effective, but what does it tell us about ourselves and how
> we must live? The brief answer to this is: nothing. Science has always
> worked assiduously to avoid being a religion, faith or morality.[45]

According to Appleyard, this corrosion by science of ancient beliefs, on
which good personal morals and good social practices were based, is a result in
part of a tactical mistake by the Christian church. This can be traced back to the
attempt by Saint Thomas Aquinas in his 1266 *Summa Theologica* to unite
"Christian and Classical wisdom" and, in particular, to make Ptolemaic astronomy the center of religious cosmogony. In eventually popularizing the Copernican view of the Solar System, Galileo (who, when Appleyard was writing, was
still considered a guilty man by the Catholic church) "achieved everything and
nothing" since his science led to the conclusion that "there is nothing especially
privileged about our position" in the cosmos. In moving the center of the Solar
System from the Earth to the Sun, Galileo had "cut scientific man adrift from
his moorings." "At the heart of this spiritual problem lies the lack of self," Appleyard explains. In the centuries following Galileo, science has progressively
become the culture of the entire world—"now its ambition is to unravel the
workings of the human self." The message of the book is twofold: that "liberal-
democratic society has been created by the scientific method, insight and belief" and that "both this society and science are inadequate as explanations and
guides for the human life."

Appleyard is especially hostile to attempts to make science publicly accessible and thus to the whole movement for public understanding of science. "The
popularizers are popular because they make sense of all the difficult science"—or
so they claim. In fact, what they are offering is a trap, "a hollow mechanistic
vision . . . the reduced version of ourselves."

Many of the views expressed so pungently by Appleyard are themselves echoes of the views of the 19th-century poet and literary critic Matthew Arnold. He scorned the sort of knowledge that science and scientists could provide as being simply "fact upon fact," lacking the ability to relate to the "powers" that go to make up and enlighten human life.[46] The differences between knowledge as information and knowledge as wisdom, particularly in relation to scientific knowledge, have also been elegantly explored by the English philosopher Mary Midgley. She doubts the adequacy of science alone to answer the "big questions," asking: "Is it actually plausible that the particular kind of curiosity that impels people to study the natural sciences is the one central demand of human nature?" Modern science has led to the intellectual scene being "mapped into ever smaller and more discrete provinces." This has resulted in a failure to concentrate on the "province of thought" that deals with "what matters" rather than trivial pursuits such as the "sand-counting" of daily scientific research. In Midgley's view:

> A remarkable attempt has been made in this century to withdraw the notion of knowledge from this province of thought [of what matters], indeed to cut it off radically from all the rest of life. . . . When knowledge is secluded in this way and equated with information, understanding is pushed into the background and the notion of wisdom is quite forgotten.[47]

One answer to those like Appleyard—whose message might be interpreted as a desire to return, intellectually at least, to pre-Enlightenment days—has been to meet them head on. Writing in his last book before his death in 1996, Carl Sagan, astronomer and champion of the search for extraterrestrial intelligence, deals with the "God hypothesis"—the idea that there is a supreme extraterrestrial—even-handedly. "The fact that so little of the findings of modern science is prefigured in Scripture to my mind casts further doubt on its divine inspiration. But of course, I might be wrong."[48] Others are far less unsure. Chemist Peter Atkins, for example, is proud of the "spiritually corrosive" nature of science:

> Our view is that modern science is indeed striking at the concept of the soul and the existence of God. I do not consider it irksome that science is nibbling at the roots of belief. If individuals are to be governed, better on the basis of truth than invention and falsehood.

Against those arguing that science is unsatisfying and robs the world of its natural beauty, he rejoins:

> I do not consider that knowing the origin of the world, and accepting its lack of cosmic purpose, in any sense diminishes my delight. Sci-

ence . . . respects the power of human understanding far more than any religion ever will.[49]

## WHAT DOES THIS ALL MEAN FOR THE PUBLIC UNDERSTANDING OF SCIENCE?

In this world, nothing stands still. That is certainly true for the issues discussed in this chapter. At the time of writing, battles are still being won and lost on all sides of the divides.

In the summer of 1995, Paul Gross and Norman Levitt assembled a major conference under the auspices of the New York Academy of Sciences to carry on the work they started in *Higher Superstition*, although on this occasion the "religious right" came under fire as well as the "academic left."[50] In 1996, New York physics professor Alan Sokal used a novel way to expose what he saw as "an apparent decline in the standards of rigor in certain precincts of the academic humanities" by first successfully submitting a "spoof" article on the "transformative hermeneutics of quantum gravity" to the postmodernist journal *Social Text* and then writing an admission that his article was a "parody."[51] Sokal's success in persuading *Social Text*'s editors to publish his article without first checking with a physicist to see if his science was valid (it was, deliberately, complete balderdash) was seen either as a brilliant tactic to expose the shallowness of current humanist commentaries or as a heinous breach of academic trust, depending from which side of the "science wars" the comment came. Edinburgh philosopher David Bloor, along with colleagues Barry Barnes and John Henry, restated the principles of the "strong program."[52] Sociologists Harry Collins and Trevor Pinch slugged it out with Cornell University physicist David Mermin over their interpretation—as portrayed in *The Golem*—of the experiments that "confirmed" Einstein's theory of relativity.[53] But there were also signs of reconciliation in the academic community; in July 1997, Collins hosted an international conference at the University of Southampton, and both Mermin and fellow physicist Kurt Gottfried—another of *The Golem*'s most strenuous critics—attended; there was hardly a cross word all day.[54] Later that summer, sociologist of science Michael Lynch told the British Association for the Advancement of Science that "this was not a war. No-one got killed, and no civilians were caught in the crossfire."[55]

However, as we have seen, those fighting the cause of science, particularly in the United States, portrayed the public—the "civilians"—as misguided, misled victims of the anti-science onslaught. The sociologists and historians—very particular publics for science—may have revealed their own understandings of science, but were the general public actually at risk? Have astrology and postmodernism really turned lay people against science? Or were scientists' expressed concerns about the public mere devices to deflect attention from the scientific community's own unease in the face of the criticisms being made of science? The debate continues.

## Chapter 4

# Popularization, Public Understanding, and the Public Sphere

In discussions of science, popularization, and the public, talk is frequently of activity—of what one should do in order to achieve better or more public understanding of science. Rather less attention has been devoted to articulating the philosophies and models that inform and drive popularizing activity. The communication process, the functions—intended and otherwise—of popularization, conceptualizations of the public as potential recipients of scientific information, and, indeed, the notion of public understanding of science itself are just some of the complexities that are, perhaps too readily, taken for granted. This chapter looks behind the activity to find the models and motives of science in public and assesses the work of researchers who aim to understand the public understanding of science.

## POPULARIZATION: WHY AND FOR WHOM?

As Chapter 2 indicates, the relationship between science and the public has been a long one. The attitude of the scientific community toward popularization has varied widely and dramatically, both over time and between disciplines; according to Anne Branscomb of the American Bar Association, writing in 1981:

> Our society has not determined whether it is the responsibility of the professional scientists to communicate only to their peers, or also to elected or appointed government officials, or directly to the public.[1]

The rise of the "public understanding of science" movement of the 1980s brought an expression of positive attitudes toward popularizing activity by scientists,[2] but despite high-level approval within the scientific community, popularization is still not seen by many academic institutions as something to be encouraged or rewarded; popularizations often do not count as publications in academic audits, for example.

The changing motives of popularizers and the oscillating attitudes of the scientific community have left a legacy of confusion and ambivalence: scientists raised in one tradition or in a particular political or social environment may spend most of their careers in circumstances where different rules apply. Popularization has been used to empower the workers and to keep them in their place; it has made claims for the privileged status of science, for example with regard to funding, and has also been suppressed in order to maintain science's privileged status. Popularization exaggerates and highlights tensions in the scientific self-image: science is neutral but concerned, commonsense but special, democratic but authoritative.[3] Scientists might speak to a press officer but not to a journalist; they might both collaborate in media productions and condemn the methods and output of media professionals.

According to communications scholar Rae Goodell, the scientific community exercises a "powerful system of social control" over its members who popularize. According to scientific tradition, the scientist-popularizer should follow certain rules, among which are that one should popularize only when one's productive research life is over; stick strictly to a specific area of expertise; act only to improve the public image of science, especially where funding may result; and avoid extremes of opinion. Another unwritten rule of science popularization is that a scientist must first establish a reputation as a credible researcher before he or she is entitled to communicate with the public.[4] Yet another rule is that the popularization of a particular piece of knowledge should happen only after "conceptualization" and "documentation"—the scientist should have developed the knowledge and published it in the technical literature before presenting it to the public.[5] Many scientists have fallen foul of these strictures, not least since these are largely at odds with the demands of the mass media (see Chapter 5).

Science communication scholar Bruce Lewenstein suggests that the rules of appropriate behavior with regard to popularization are used self-servingly—they are stressed by scientists who want to criticize or limit other scientists' behavior but are ignored by the same scientists with regard to their own behavior.[6] Scientists who do not popularize tend to see popularization as something that would damage their own career; however, they also think that other scientists use popularization to advance their careers. Popularization does seem to magnify career successes and failures, particularly in fields where popularization has a strong tradition. It has been argued, for example, that Stephen Hawking's success as a popularizer has inflated his standing within science.[7] On the other hand, some scientists associate popularization with either willfully or inadvertently wrong science, so that any science in the public domain is "taint[ed] or . . . intellectually suspect."[8]

In Goodell's classic study of popularizers, her "visible scientists," among them the astronomer Carl Sagan, the biochemist Paul Ehrlich, and the anthropologist Margaret Mead, reported no adverse effects on their scientific careers from their popularizing; but they all held tenured posts before they became visible, and their visibility brought them money. As Ehrlich said: "I don't know how anybody could affect my career in any way that would really disturb me." Mead reported that the penalties were more social and personal—snide remarks and snubs—than institutional or professional. Goodell's visible scientists share certain characteristics: they have a "hot topic"; they are "prepared to take an unqualified stand"; they are "mavericks, frequently at odds with fellow professionals, the scientific establishment, the political system, religious tenets..."; they are "controversy prone" and "revolutionary" in their scientific work; they are articulate and colorful; and they are visible within their field before they become visible to the public or are associated with a high-profile institution. These are characteristics required by media that emphasize news value and the reliability of sources.[9]

But the contradictions and tensions in the scientific community's attitudes to popularization remain. Goodell reported that "visible" scientists can be rejected by their colleagues irrespective of good practice or of the validity of their ideas; and even successful, persistent popularizers are critical of the mass media. According to Goodell: "Scientists exchange horror stories about the press the way laymen discuss their operation scars."[10] Her visible scientists, despite—or perhaps because of—their long experience with the media, still complain of sensationalism, inaccuracy, and oversimplification.

The mass media serve many of the same purposes for scientists as they serve for other individuals—the promotion of a personality, idea, or cause, a few moments of fame, money, and entertainment can all be achieved through media activity. Some scientists have personal reasons for popularizing. According to T. H. Huxley:

> . . . some experience of popular lecturing had convinced me that the
> necessity of making things plain to uninstructed people was one of the

very best means of clearing up the obscure corners of one's own mind.[11]

According to Albert Einstein, popularization has a spiritual function:

> Restricting the body of knowledge to a small group deadens the philosophical spirit and leads to spiritual poverty.[12]

Scientist-popularizers take pleasure and a feeling of a duty fulfilled from the altruistic act of sharing their knowledge: despite its public relations function, popularization has retained its historical association with a liberal agenda. According to philosopher I.C. Jarvie, "media popularizers seem to be the last representatives of the philosophy of science of the Enlightenment movement."[13]

Sociologist of science Harry Collins has suggested that informal communications are as much a part of scientific life and practice as formal communications,[14] and historian of science Martin Rudwick advises against thinking of sources of information about scientists' thoughts as divisible into distinct categories of private and public. He suggests instead a "continuum of relative privacy":

> No single activity and no single class of documents gives privileged or even preferential access to the scientist's "real" thoughts or beliefs, since all alike are the embodiments of modes of thinking and doing that form essential components of the whole process of scientific work.[15]

Popularization is part of the making of scientific knowledge, as well as of the sharing of it.[16] Thus, scientists who popularize are doing science in public. However, the science that is "done" in this sense is unlike the science done within technical media. The rhetorical traditions of academic science serve particular purposes for the scientific community: the passive voice and the methods/results/conclusions formalism that identify scientific writing and proclaim both objectivity and hypothetico-deductivism—science's special province—disappear in popular communications, which therefore describe a very different kind of science (see Chapter 5).[17] For these and other reasons, and no matter how entwined the popular and the technical might appear to disinterested observers, the scientific community has long maintained the separateness of the two forms of communication, and of the two communities they reach.[18] Indeed, popularization is itself a means to maintain these separations: according to Jarvie,

> . . . the scientific community employs various communication processes and structures in a strategic manner that help the community preserve the privileged status of scientific knowledge in . . . culture.[19]

The popularization of science has repeatedly become an issue of social and political concern, both to scientists and to others; and the purposes to which different groups put popularization has been the subject of academic investigation. According to Lievrouw:

> Popularization is . . . essentially a communication process that facilitates the gathering of resources for pursuing certain lines of research.[20]

"Resources" here could mean political patronage (for a new supercollider) or public approval (for genetically engineered food). Popularization is essentially an act of persuasion:

> The purpose of communication is to reconstruct one person's idea in another person's mind.[21]

Sociologist Stephen Hilgartner has argued that scientists use popular accounts for social and political purposes within the scientific community as well as to reach patrons and the public; thus, potential audiences for popularizations include other scientists, and popularization is a means of internal and interdisciplinary communication.[22] Communication with closer colleagues via the public media can reinforce allegiances and convert opponents.[23] Converting opponents is a difficult task—which may be why popular media are used for this purpose: they allow for simplified accounts which might be more easily understood and therefore likely to attract more followers.[24] Historian Marcel LaFollette offers as examples of this strategic use of popularization the approaches made to the public in the early years of the 20th century by eugenicists when they failed to establish a scientific constituency and the response in the popular media, between the wars, of psychologists to academic attacks on the legitimacy of their discipline. The aim of these communications was to reach not only the public but also, and importantly, the scientific community. Thus, popularization, if it is persuasive, can be a means to power; however, if all popularization is "tainted," then it can also be a way of losing credibility.

Scientists often turn to popularization when their institutional links are weak, for example when they retire or when they support unpopular ideas. Lewenstein has suggested that scientists use popularization to resolve conceptual controversies because popular media communicate more quickly (if more ephemerally) than technical media.[25]

In a field that is poorly institutionalized (perhaps because it is new, or because of rapid or radical conceptual change), security and rewards are hard to find, and scientists may look to popularization to provide these. In the early stages of the life of a discipline or field, popularization may be easier because the scientist tends to be working directly with phenomena and with empirical data[26]; at later stages, when the science has moved away from phenomena and toward theory, popularization may be more of a challenge. Popularization is used to define the boundaries of a new field—it stakes a claim on a new domain. When a debate occurs in a field which, though established, is not fully institutionalized, some communication of ideas must inevitably take place beyond institutional boundaries. Such popularizations by scientists announce to the public that certain ideas are now within the province of science. A recent example might be the plethora of popular works on consciousness that were published in the mid-1990s. These books and articles were written by people from a variety of disciplines (electronic engineering, molecular biology, psychology, mathematics); a review on the cover of Daniel C. Dennett's *Consciousness Explained* says that it "will set the agenda," the implication being that there is much work still to be done.[27] This is popularization that merits communications scholar Baudouin Jurdant's description of "autobiography": it is science defining itself in public.[28] It is also popularization that emphasizes the difference between professional and lay communities, rather than bringing the two closer together:[29] Consciousness, like cosmology before it, may once have been the subject of popular myth and idle ponderings, but now it is the stuff of rigorous science.

## MODELS OF COMMUNICATION

Simple linear models of the communication process are common in the science communication literature.[30] Based on mechanical "sender–transmitter–receiver" communication models developed by electronic engineers in the 1940s, they place science and the public at opposite ends of a trajectory, with journalists and other mediators sometimes somewhere in between.[31] In communications scholar Christopher Dornan's account:

> Science is seen as an avenue of access to assured findings, and scientists—in the dissemination of these findings—as the initial sources. The members of the laity are understood purely as recipients of this information. Journalists and public relations personnel are viewed as intermediaries through which the scientific findings filter. The task for science communication is to transmit as much information as possible with maximum fidelity.[32]

According to Hilgartner, scientists see popularizations as inferior to technical accounts,[33] so while communications flow in both directions, the "public" end is "downstream" and the "scientific" end is "upstream." One striking articulation of this communication hierarchy describes five levels of audience for communications: the highest is "the scholar"; the lowest is "the man in the street."[34] The model also allows scientists to see popular communication as entirely separate from scientific communication. It is, according to Hilgartner, the "dominant model" of popularization—not because it necessarily describes how popularization actually works, but because most popularizers tend to behave as though they believe this is how it works. Scientists, at the top of the knowledge heap, communicate valuably with each other through technical media and might occasionally aim a few scraps, in the form of popularizations, at the humble and passively receptive public down below.

The linear model is a one-dimensional version of the two-dimensional diffusion model of popularization, in which the scientific community emits information to be received by the public. Communications scholar Robert Logan argues that the diffusion model is essentially a model in which science communication is intended to be purely persuasive (one might say propagandist); his study of health communication concluded that such communication does little for the public's health literacy.[35] The diffusion model, like the dominant linear model, describes what sociologist Brian Wynne has called "top down" communication, in which information that is meaningful in the scientific context in which it originated is likely to arrive in an entirely different lay or public context, in which people are unlikely to accommodate it (should they choose to) without substantial interpretation and adjustment.[36]

While linear and diffusion models have some explanatory power on a gross level, they contribute little to an understanding of the complex communication systems in which scientists routinely engage. For example, locating the "scientific" and the "popular" at opposite ends of a scale is difficult to maintain in practice. If a microbiologist reads a newspaper feature article about food poisoning, does the article become a technical communication? If a sick layperson were to read up on the latest research into his condition, are those academic papers now popularizations? When a scientist first notices a phenomenon and tells a colleague, is she popularizing? Is a funding application a popularization for a lay readership? What about a forensic scientist's statement in court? Hilgartner suggests that rather than try to separate the popular from the technical, we can think of a continuum of communications on the linear scale.[37]

A model that offers a more integrated approach to communications and their interactions without privileging any medium is the "web" model suggested by Lewenstein, in which technical and popular media (however defined) interact in complex ways, informing and referring to each other. According to Lewenstein, "analysing the role of the mass media . . . must be an exploration of the complexity of interactions among all media."[38] Lewenstein's web model sits easily with the idea of communication networks described by sociologist Niklas Luhmann. A communication network model might consider the boundaries of science not as fences between separate domains of cultural or intellectual activity, but as limits of open territories which may overlap with other domains, and which are themselves superposed on culture as a whole. These territories contain communities that define themselves by communication.[39] Thus, a community is defined by communication between its members, and membership in the community is contingent on the sending and receiving of communications.[40] In this model, the popular and the scientific become part of the same communication system, and the communicators involved become part of a single community. In line with arguments from the sociology of science, a communication network model allows popularization to be considered not as something peripheral to scientific activity, or as deviant or pathological, but as an integral function of normal scientific life.

## UNDERSTANDING THE PUBLIC UNDERSTANDING OF SCIENCE

The popularization of science is seen as one route to public understanding of science. But what constitutes an understanding of science or a reasonable level of scientific literacy? Many people have attempted to answer this question, and the result of their efforts has been a variety of prescriptions of what the public ought to know.

Scientists' own aims in the public communication of science—or rather the public understanding of science, which is the supposed result of public communication—have been expressed consistently for some time.[41] As discussed in Chapter 1, these arguments all stress the positive side of science; they are arguments not so much for the public understanding of science, but, as Lewenstein has pointed out, for the public appreciation of science: the masses shall know about science and will therefore will value it.[42] Yet what the public knows and understands and how this might relate to their attitudes to science are difficult phenomena to access. The much wielded but in some ways rather blunt instrument, the quantitative social survey, has produced a particular representation of the public understanding of science. Such surveys have been run regularly on behalf of the National Science Foundation in the United States since 1972 and tend to show that—in that country at least—levels of scientific literacy are consistently low. Another consistent result from these surveys is that people's self-reported interest in science is much greater than their self-reported informedness about science. On the face of it, this lends weight to arguments in favor of placing more scientific information in the public domain to meet the reported demand. But how far this self-reported and apparently unsatisfied interest translates into a burning desire to remedy one's lack of knowledge is quite another matter.

In terms of levels of scientific knowledge, as in the United States, in the United Kingdom a survey fielded in 1988 revealed that fewer people met the criteria for being "scientifically literate" than considered themselves to be informed about science.[43] (Eight years later, the situation had changed little.) However, one problem in deriving the percentage of scientifically literate members of the public arises from the different ways in which questions can be formulated. For example, the open and abstract question "What does it mean to study something scientifically?" floored all but 3 percent, on the questionnaire's criteria. But more than half knew how to carry out a successful drug evaluation when given a choice of answers; closed and concrete questions are much closer to everyday experience than abstractions about the "scientific method." Thus, as social psychologists Martin Bauer and Ingrid Schoon have pointed out, measures of public understanding of science really tend to measure the extent to which the public think like scientists, which, since they are not trained scientists, is, not surprisingly, not much.[44] This difference between scientists' and public thinking is interpreted as public ignorance, and it is this ignorance that these surveys highlight (surveys could be fielded that would just as efficiently show public ignorance of economics, history, literature, horticulture, architecture, etc.). These surveys offer rather less insight into the extent to which, or the way in which, the public understands science and give weight to a "deficit model" of public understanding of science which conceptualizes the lay mind as an empty bucket into which the facts of science can and should be poured. Critics of the deficit model point out that it follows from a prescriptive, top-down view of public understanding of science, in which the promotion of science is a major concern. Like the dominant model of popularization, the deficit model locates knowledge and expertise solely with the scientists and keeps them at the top of the heap.

Professor of public understanding of science John Durant has examined various approaches to scientific literacy—including that implicit in the deficit model—and has classified them under three headings: understanding as knowing a lot of science; understanding as knowing how science works; and understanding as knowing how science *really* works.[45]

Many in the scientific establishment consider that a public-understanding-of-science deficit could surely be remedied by the liberal application of scientific knowledge. But Durant argues that knowing a lot of science is clearly not the same as understanding science. While facts may be interesting and no bad thing in themselves, knowledge of facts does not imply an understanding of their significance or implications, nor of their place in the wider scheme of science. More important, knowing the facts is often little help to citizens who are trying to come to terms with contemporary issues in science. The important issues raised by scientific and technological innovations, for example, concern new science, where the facts are often disputed and are usually inaccessible in practice.

Nor is understanding how science works necessarily helpful, at least if standard textbook accounts of "the scientific method" are used to explain the workings of science. Philosopher Alan Chalmers sums up these accounts as saying:

> Scientific knowledge is proven knowledge. Scientific theories are derived in some rigorous way from the facts of experience acquired by observation and experiment. . . . Science is objective.[46]

This notion that there is a "scientific method" is, according to chemistry professor Henry Bauer, an obstacle in the way of achieving better public understanding of science.[47] Bauer argues that the "central fallacy" in many efforts toward public understanding of science is the idea "that there exists an entity called 'science' about which sweeping generalizations can validly be made." The notion of the "hypothesis–test–verification/falsification" strategy of research is, according to Bauer, "a myth." In reality, the practice of science consists of a series of somewhat ill-defined puzzle-solving exercises, which may or may not be carried out cooperatively, the results of which exercises then pass through a system of more or less effective filters, including a "reality therapy" check against nature, before they are (often provisionally) accepted as being true. However, the public should be aware that "to say that the scientific method is a myth is to say that it is not literally true, which is not to say that it is worthless." With an upbeat flourish, Bauer concludes:

> . . . humankind can take collective pride from the persistent determination to submit to reality therapy that has produced not only the science that we now know but also an understanding of how to go about learning more.[48]

As we saw in Chapter 3, the standard account of the scientific method does not bear much resemblance to what sociologists, historians, and philosophers understand as the real practice of science either. From the public's perspective, this account fails as soon as scientists disagree in public. When crises erupt in the public sphere—new diseases, food scares, environmental problems—the public is invariably confronted with scientific experts who draw very different conclusions from the same "facts."

Durant also points out that many pseudo- or parascientists use scientific methods to produce their effects or arrive at their predictions. How, then, can the public decide which of the conflicting but apparently scientific claims to believe? Durant suggests that understanding how science, as a practice and as an institution, really works to generate knowledge might help here. Sociologists of science have suggested that this understanding can be achieved by looking critically at individual cases of how scientific investigations are conducted and how scientific decisions are made. Such case studies have included how scientists came to believe in Einstein's theory of relativity and why they lost interest in whether or not memory was an edible commodity.[49] The problem with this analysis of understanding science, however, is that in order to understand how science really works, one also needs to understand quite a lot of science.

Durant suggests that understanding science as an institution could help the public to differentiate between reliable and unreliable knowledge: that it is the professional aspects of science, such as the training of scientists, research protocols, peer review, and the continuous scrutiny to which scientific knowledge is subject within the scientific community, that generate, in the end, reliable knowledge. Public issues tend to arise from new science—from science-in-the-making. In these circumstances, what the public sees is not reliable knowledge, but reliable knowledge in the process of being made, with all the uncertainties and conflicts that inevitably arise while scientific ideas are being turned into reliable knowledge. According to Durant:

> . . . the public needs more than mere factual knowledge . . . ; and it needs more than idealistic images of "the scientific attitude" and "the scientific method." What it needs, surely, is a feel for the way that the social system of science actually works to deliver what is usually reliable knowledge about the natural world.[50]

We test these prescriptions for public understanding of science on a real example of science-in-the-making in Chapter 6. But however one wishes to define scientific literacy, there is a more fundamental question: with increased efforts, or better communication techniques, can we look forward to a society in which everyone is "scientifically literate"? Or is science, at least in its modern form, just too difficult, just too odd, for laypeople to grasp?

According to Morris Shamos, a physicist and long-time participant in the debates about science education in the United States, the prospects for public understanding of science are bleak. For Shamos, it is not the scientific method that is the myth, but the whole concept of scientific literacy.[51] He points out that although administrations of all political persuasions have put billions of dollars into various science education projects and initiatives since World War II, American citizens are still woefully ignorant of science. In his definition, a truly scientifically literate person need not "have at his or her fingertips a wealth of facts, laws or theories" but should be capable of explanations about the natural world "not simply by rote, but with the ability to extrapolate to other related examples." "Nevertheless," Shamos concludes, "even this modest criterion puts scientific literacy beyond the reach of most educated individuals." Thus, "the notion of developing a significant scientific literacy in the general public . . . is little more than a romantic idea."[52]

One strategy Shamos suggests is to give up altogether on the attempt to bring all American citizens to an elementary understanding of the content of science. Much better to concentrate almost all the available resources on the 20 percent of school students who show most aptitude for science and get them to a reasonably proficient level. The "20 percent solution" is significant because it might allow society as a whole to function at a workable level of scientifically literacy. For example, there would be a 93 percent probability that at least one member of a jury, faced with forensic evidence, would be "scientifically literate" and a 73 percent chance that this person would be joined by an equally accomplished colleague.[53] For the rest of the citizenry, Shamos sets much lower goals than, say, an appreciation of the significance of the genetic fingerprinting of a sample of dried blood on a discarded glove.

> What we seek is a society that (a) is aware of how and why the scientific enterprise works and its own role in that activity, and (b) feels more comfortable than it presently does with science and technology.

Shamos also favors the use of neutral "science courts"—making use of scientists to assess the claims for, say, safety and environmental friendliness of new products on behalf of their less scientifically knowledgeable peers—and a "science watch" of ordinary citizens to ensure that science genuinely serves the public. That means concentrating the education of those outside the chosen 20 percent on an appreciation of "the cultural imperative of science," the "practical imperative" concerning health, safety, and the environment, and the "proper use" of scientific expertise. The education system that Shamos therefore envisages would have one-fifth of the class going off to the physics, chemistry, and biology laboratories while the remaining students took classes in science and technology studies.

Shamos's argument is a pragmatic one, a question of practicalities and social priorities. It might be countered by saying that the issue of scientific literacy is, in fact, a matter of money and resources and that the problem is that the amounts devoted to its realization to date, while admittedly very considerable, are still simply not enough. With more, more could be achieved. But what if there is something intrinsic to modern science that makes the goal of enabling all the public to understand science essentially impossible? What if scientists, either by nature or nurture, really are a breed apart? In the last century, the view of T. H. Huxley was that science was no more than consistently applied common sense. All people had to do was to stick to their guns, take careful note of what they were observing via their everyday senses, and follow through the logic of their arguments and they could do science just as well as he could. In 1953, physicist J. Robert Oppenheimer took a rather different view: he explained that you could not expect everyday concepts to carry on working normally once you entered, for example, the strange world of quantum particles; you needed "uncommon sense." Oppenheimer told his audience:

> Common sense is not wrong in the view that it is meaningful, appropriate and necessary to talk about the large objects of our daily experience . . . [it] is wrong only if it insists that what is familiar must reappear in what is unfamiliar.[54]

But Oppenheimer still considered it a very worthwhile enterprise to try to familiarize his listeners with that strange world and drew heavily on everyday analogies to do so, before describing the pleasure that even the casual caller might obtain from a visit to the "many-roomed House of Science."

More recently, questions have been raised as to whether even views on science and common sense as cautious as Oppenheimer's are sufficiently realistic. Embryologist Lewis Wolpert has tackled the issue of what he calls the "unnatural" nature of science, in which he contends "if something fits in with common sense it almost certainly is not science." He goes on:

> One of the strongest arguments for the distance between common sense and science is that the whole of science is totally irrelevant to people's day-to-day lives.[55]

One reason for this, says Wolpert, is that "science is not just about accounting for the unfamiliar in terms of the familiar." Quite the contrary: "Science explains the familiar in terms of the unfamiliar." An everyday example can show what Wolpert means. In 1993, the fast-food chain McDonald's was sued in the United States by a woman who spilled hot McDonald's coffee on her lap while driving. Her argument was that the coffee was "too hot"—on the face of it, a straightforward complaint. However, a scientist looking at that argument would say that the "hotness" or temperature of the coffee was not the point. After all, many children have held sparklers and had the flying fragments of white hot metal land on their bare skin with no ill effects. What counts is not the temperature but the overall heat capacity. Furthermore, there are also considerations such as the rate at which the coffee soaked through the woman's clothing, the conductivity of her skin when the hot liquid reached it, and any number of other factors to take into account. By the time the scientists are finished—including those wanting to invoke kinetic theory and the first law of thermodynamics to explain what is happening—we have certainly left the familiar realm of crying over spilt coffee and are on fairly unfamiliar ground.

The sort of thinking and conceptualizing involved in a scientific analysis of the familiar, argues Wolpert, is of an essentially counter-commonsense kind. His argument also makes great play of the historically accidental nature of science—science in this case severely restricted to exclude technology. "The development of Western science has been based on two great achievements, the invention of the formal logical system by the Greek philosophers and the [Renaissance] discovery of the possibility of finding out causal relationships by systematic experiment," claims Wolpert. In the sixth century B.C. Thales of Miletus postulated that the universe was water. "Never before had someone put forward general ideas about the nature of the world . . . . For the first time . . . there were laws controlling nature . . . and these laws were discoverable." Not only that, but when Anaximander proposed that the universe was fire, he and Thales could not both be right—theories were in competition, they could not just coexist. Galileo's breakthrough was to challenge the mechanics of Aristotle by using experiments, even if many of them were of the thought variety. Science, by these two great developments of thinking, separated itself from the everyday.[56]

Some of Wolpert's critics pointed to the inherent elitism in his position. "Once science is ethereal, individual, the stuff of far-sighted leaps and bounds, what role for lesser mortals? To clap from the sidelines?" historian of science Michael Shortland demanded to know.[57] "There are scientists and the rest. If you adopt this position . . . public understanding of science would reduce to understanding the superiority of the scientific mode of investigation," added science communicator Jon Turney. Other critics focused on the way in which what today is front-line science becomes common sense tomorrow. "Science is not intrinsically counter-commonsensical; on the contrary it is the major force for change in common sense," rejoined sociologist of science Harry Collins. Nevertheless, Wolpert's arguments still pose problems for those trying to get the public to understand science as it is developing now, as in the case, say, of genetic technologies, with all their potential and implications. Wolpert's own argument—that you could get a feel for the whole of science if you had at least done a bit of it—is a strong one for encouraging hands-on experience in both the formal education system and in public expositions.

## UNDERSTANDING THE PUBLIC

At the heart of all these debates are the public, who present a major challenge to those practitioners and analysts who want to know just for whom their efforts are being made, and what those people are supposed to gain from, or do with, the information or understanding they receive. Different conceptions of "the public" lead to different strategies for public understanding of science, just as different conceptions of "understanding" lead to different assessments of the efficacy of the strategies. Often, though, both the public and their understanding are black-boxed to alleviate the need to decipher them, or they are accessed through anecdote and personal experience: one's neighbor or aunt becomes the paradigm for the public, and how they think becomes the paradigm for public understanding. These are sensible moves, perhaps, given both the complexity of the entities involved and the need for popularizers in the real world to get on with the job at hand. However, "the public" as a body has long been the subject of sociological inquiry, and some ideas have emerged from these efforts that cast some light into the black box and that lift our understanding of the public understanding of science beyond our immediate experience.

In everyday language, the word "public" is often accompanied by positive, active connotations, which recall the idealized era of Athenian democratic citizenry, keenly debating the issues of the day and arriving at decisions by force of democratic argument. So we hear of someone being "public-spirited"; they are carrying out "public duties" in the "public interest," no doubt listening carefully to "public opinion," and offering due "public accountability." And it is important that the decisions reached by those in "public life" are open to "public scrutiny." But, according to the sociologist Jürgen Habermas, this active "public sphere" is a relatively recent development, intimately bound up with the development of capitalist, industrial society with its roots in the 18th century—the period of the Enlightenment.[58]

Prior to the development of bourgeois, capitalist society, Habermas says, authority was maintained by shows of "publicity" as much as shows of force. In feudal society, courts regal and ducal would travel their domains, stopping off here and there so that the personages of power could be represented to their subjects.

"The staging of publicity involved in representation was wedded to personal attributes such as insignia, dress, demeanour and rhetoric—in a word, to a strict code of 'noble' conduct," Habermas explains. But as the feudal order began to crumble, in the wake of the English Civil War and during the build-up to the French Revolution, authority was increasingly confronted rather than accepted. "The medium of this confrontation was peculiar and without historic precedent: people's public use of their reason." The outcome of this bringing together in public of the critical facilities of private individuals, activity associated with the coffeehouses of London and the salons of Paris, was the formation of the public sphere:

> The bourgeois public sphere may be conceived above all as the sphere of private people come together as a public; they soon claimed the public sphere regulated from above against the public authorities themselves, to engage them in a debate over the general rules governing relations in the basically privatized but publicly relevant sphere of commodity exchange and social labour.[59]

The notion of public here, then, is of active, independent, and responsible citizens, with power, wealth, and influence, armed with the latest information and debating the conditions of their social existence and interests. A consensus for the common good emerges from the battle of individual wits. Many proponents of the public understanding of science see their activity in that light, supplying relevant, often vital information so that informed citizens can discuss matters in which science plays a part and arrive at a democratic consensus on the way forward. Hence, a prominent entry in the list of "benefits" to be derived from the public understanding of science that we outlined in Chapter 1 is the advantage to democratic society, as well as to individuals, of a scientifically knowledgable and empowered public.

But alongside the rather worthy attributes associated with "public" understanding activity are the less highly regarded ones of "mass" communication. According to mass communications scholar Denis McQuail, the concept of "mass" carries with it connotations of large numbers of people, rather widely dispersed and heterogeneous to the point of having little in common—thus hardly interacting with one another at all, other than via their common point of contact with a media communication (be it the *Washington Post* or a television soap opera).[60] Anything produced for "mass consumption" is something rather to be looked down on, it may even produce "mass hysteria," and "mass movements" are to be feared. Mass culture is dished out via the mass media to a diffuse, disorganized mass audience that is more susceptible to emotion than to reason. To complete the classical metaphor, the mass audience is reminiscent of the Roman "lumpenprole-tariat," idle, unproductive, and kept in place by bribes of bread and circuses.

In Habermas' view this is the situation we have reached today with the development of the mass media, particularly television:

> In comparison with printed communications, the programs sent by the new media curtail the reactions of their recipients in a peculiar way. They draw the eyes and ears of the public under their spell but at the same time . . . place it under "tutelage," which is to say deprive it of the opportunity to say something and to disagree.[61]

This process, according to Habermas, has led to a "refeudalization" of the public sphere, where even public debates about crucial policy questions are reduced to the level of entertainment.

Many proponents of the public understanding of science are concerned that the various scientific establishments have been too much involved with this mesmerizing aspect of public understanding—in which the masses are awed into passivity and unquestioning support for science by the sheer brilliance of modern research—and too little involved with the active, empowering side of the project. The top-down deficit model of public understanding of science, which is generally followed by scientists, is further criticized for its naive conception of the public. It views them as passive recipients of information, taking no account of how the information they receive will interact with their pre-existing knowledge and attitudes and ignoring any demands they may have for what they learn to be relevant to their individual situations. What is required, it is argued, is for scientists to view the public, not as a lumpen mass, but as specific groups of active and thoughtful citizens.

However, conceptualizing the public remains a challenge: it requires a degree of reflexivity and yet at the same time a degree of objectivity, for "the public" is not only ourselves but also, sometimes, not us; it includes, sometimes, the scientists who do science, and the institutions of science, and also the people who work in the media, and media institutions. In discussion of public understanding of science, the field is narrowed: the "public" becomes the "lay public"; but again the question of who counts as a layperson is not a simple one. Most physicists, for example, are not embarrassed to admit that they know little biology: when it comes to biology, physicists are laypeople—they often know no more about biology than any other member of the public. The same can be said about different fields within physics: a nuclear physicist may happily know little about electronics. So within science, and even within fields within science, the dividing line between "expert" and "layperson" is flexible and dynamic, and the same goes for society as a whole. Looking at people in general, we can say that everyone is an expert in a few fields and a layperson in an infinite number of others. Most people have some expertise in fields which are close to them: for example, the geography of where they live, the habits of their pet, or the minutiae of their job. People who live somewhere else, have a different pet, and do a different job will have different expertise. These two sets of knowledge are equivalent: each is valuable in its own context and of little value outside of it.

There is much evidence to suggest that people acquire, or are prepared to accept, only that scientific information which they need for their own particular circumstances, and little more.[62] Even people working in highly complex (and potentially dangerous) technological environments often learn only what they need to fulfill their particular responsibilities, and they trust their colleagues' expertise in the rest of the process.[63] Laypeople's understanding often remains specific to the circumstances that produced it, and is not transferred to other situations. For example, people with long-term illnesses may become expert in their particular disease, though very little of what they know may be applicable to other diseases or other people.[64] Gamblers can be adept at calculating the odds on a horse race, but they may be unable to transfer their calculating skills to an abstract problem in statistics.[65]

Thus, public knowledge, which is sometimes called "lay expertise," tends to be specific or concrete rather than general or abstract—the opposite of how scientists see their knowledge of science. But the converse is also true of science: the expertise wielded by scientists, who work with general laws, may be inappropriate in specific circumstances. After the Chernobyl disaster, farming restrictions imposed in Cumbria in northern England were based on an understanding of what scientists thought was the universal behavior of cesium in soil, but the research had been done on clay soil, whereas the Cumbrian soil is peaty.[66] Pellets issued to treat irradiated reindeer in Lapland were believed to be universally appropriate for grazing animals, but they had been designed for cows and were too big for the reindeer to swallow.[67] In both cases, local people (the lay experts) could have helped scientists plan appropriate measures, had they been asked. Instead, they felt that the "experts" were dismissive of local knowledge based on particular experience rather than on general scientific principles, to such an extent in the case of the Laplanders that they felt their whole culture to be threatened.

Studies like these have led to development of a contextual approach to public understanding of science. This calls for scientific information to be given to the public in ways that relate to their specific interests and concerns. And that, in turn, means that scientific experts who deal with the public can no longer adopt a top-down approach, giving a "generalized" public a set of facts whose importance has been decided upon in advance by the scientific great and good. Instead, it is argued, scientists should work with the particular problems and expertises of their public and tailor their communications accordingly—the task is less one of propaganda and more one of negotiation. This is clearly a more difficult task, but it is one that allows scientists and the public to work together as citizens of a scientific culture.[68] This idea of negotiation seems particularly pertinent at times of acute public concern about science: examples of such episodes are discussed in Chapter 7.

## TRUST IN THE PUBLIC SPHERE

In 1967 the atomic scientist and biophysicist Eugene Rabinowitch acknowledged the cultural and practical arguments for the popularization of science, but added another: "helping nations to appreciate the dangers science has created for their future."[69] He argued:

> To educate mankind for living in the new world created by the scientific revolution . . . this is what science popularization will have to be about for a long time. Science cannot be satisfied now with explaining to the layman the structure of the nuclear reactor or of a long-range rocket, the molecular architecture of the gene or the workings of a wonder drug. The understanding of science is now needed for a more crucial purpose than intellectual enjoyment—it is needed to teach nations how to adapt their ways of life (particularly their ways of international life) to the inescapable requirements of the atomic age.

Rabinowitch's plea is an expression of the democratic argument, but its explicit reference to the dangers of science implies that popularization is a tool that can and should be used to keep science in check. This is a rare example of a call from within the scientific community for popularization as watchdog; it reflects perhaps a point in history, and a particular community in science, where scientists were not always at ease with the results of their work.

The public too are not always at ease with the results of science. It has been argued that they fear science[70] and that they fear it because they do not understand it. Therefore, greater understanding will lead to less fear (and by implication more support).[71] This is an argument that is often put forward by industry—for example, in the early 1990s by the nuclear and biotechnology lobbies.[72] But qualitative and quantitative research has consistently demonstrated that it is lack of trust, rather than lack of understanding, that leads to fear of science and that increasing understanding is just as likely to lead to more fear as to less. Attitude research has also indicated that the public are not frightened of science in general, though they have some particular fears.[73] Nevertheless, scientists continue to use public fears about science as an argument for greater public understanding of science.

As we have seen, understanding the public is primarily a sociological, rather than a scientific, challenge. Sociologist Friedhelm Neidhardt, who has looked at the public not as a passive potential audience for science communication but as active constituents of the system in which the scientific community and its products thrive and function, has drawn some interesting conclusions regarding possible relationships and their consequences.[74] First, he confirms that it is very difficult to define the public; however, whatever scientists think the public is, they believe that the public does not trust scientists. According to Neidhardt, the public is everybody, and therefore we can only assume that it is, by and large, a lay public; and the more "public" the public is, the more "lay" we have to assume it is. For example, if "the public" at a "public" lecture consists of only Boy Scouts, or pregnant women, or senior citizens, or residents of Cedar Falls, Iowa, we might be able to make some assumptions about their interests, experiences, and expertises; but most publics—newspaper readers, voters, moviegoers—are less specific, less knowable, and therefore, we must assume, necessarily more "lay" than these.

Science communicators, particularly in the mass media, therefore have a very difficult task: what can we infer about the public when our only knowledge of them is, for example, that they own a television? The media can know very little about the cognitive or intellectual resources of their audiences, and so they use strategies that reach everyone: they get attention, arouse interest, and stir emotions—strategies which often require the extreme tactics of sensationalism, or what tends to be called, in highbrow circles, "aiming for the lowest common denominator." Neidhardt argues that factual knowledge is barely communicable, much less receivable or digestible, in these circumstances. But Neidhardt is no pessimist: he points out that what the public can do, through their engagement with the mass media and other forms of popularization, is demonstrate what is acceptable. The public can take a moral stance on science, even if it cannot understand.

Neidhardt believes that a public understanding of science is unlikely but suggests trust as an alternative. Trust, he claims, compensates on a social level for communication deficiencies on a cognitive level—it makes knowledge and understanding redundant. If trust is not forthcoming, then mistrust can be used constructively, to guide scientists toward a socially acceptable science.

Trust has been a key issue in research on public understanding of science. Many were surprised by Brian Wynne's study of apprentices working at the Sellafield nuclear power station, who, one might have assumed, would be interested in and knowledgable about the science of their workplace. But Wynne found that the apprentices were not well equipped to answer scientific questions about the physics of nuclear power generation and its risks: they did not need to be, for they trusted their employers and colleagues to provide them with a safe, functional environment in which to work. Not only that, but had they asked questions about or read up on the technicalities, they would have jeopardized the trust relationships with their colleagues that are essential if a large social institution, such as a power station, is to run smoothly and efficiently.[75]

The apprentices in the power station perhaps offer some insight into the experience of all of us as we come to terms with a scientific and technological environment. If a society depends on science and technology for its productivity—for its creation of wealth, products, food, and comfort—then that society also has to live with the side effects of science, and part of the productivity of science and technology will be dealing with those side effects. And in our attempt to know everything, we are finding out things that we cannot know for sure, and about which our knowledge is incomplete. When we want to weigh up the pros and cons of science—the science we need against the side effects we do not want—we are dealing all the time with incomplete knowledge: with what might happen, or what might result, at some time, to some degree that we cannot quite predict. According to sociologist Ulrich Beck, we are living in a risk society: a society in which all the time we are judging the benefits and risks. These might be presented to us by science in the way that medical X rays and nuclear waste are given to us by science, or they might be presented to us by science in the way that science can tell us that we are carrying a disease that might kill our children.[76] It is a feature of a scientific and technological society that many of the risks we have to deal with are scientific and technological risks. It is a feature of the separation of science from the public sphere—a separation which is both social and cognitive—that often the public's only choice is whether or not to trust the scientists. We might decide not to drink the tap water or not to live near a toxic waste site, but it is very difficult to opt out of science and technology altogether; so mostly we face the difficult problems of risk and trust.

Trust is a problem of absence or invisibility. If things are transparent to us, we do not need trust. But the public cannot see radiation pollution or the additives in food, so they have to trust (or not) the scientists to tell them whether or not they are there, and whether or not they will do them any harm. When scientists work behind closed doors, the public have to trust them (or not), because they cannot know them.

But scientists' closed doors are not usually a problem. According to sociologist Anthony Giddens, it is a feature of our culture that we ignore most of what is going on around us. We do this not out of ignorance, or out of lack of interest, but because we simply cannot take an interest in the many things that are happening in the world. The example Giddens gives is of walking along a city street.[77] We can see that other people are walking along the same street, and we make a perhaps subconscious decision not to bump into them. But, apart from that, we generally behave as though they are not there: we do not acknowledge them or even look at them. Giddens points out that these days, unlike even 50 years ago, we spend most of our time with strangers. Crammed into trains, working in big institutions, walking down busy streets—most of the people we encounter in the course of a day are strangers. We ignore the people in the street because to get by in this world of strangers, where the workings of strangers' minds are invisible to us, we have to decide to trust them. The alternative—not trusting them—is draining and upsetting. So we have developed a system of "civil inattention": we ignore the people in the street, because mostly we trust them not to hit us or steal our luggage. Once we have decided to trust them, the fact that they are strangers does not matter, because any knowledge or understanding of them becomes redundant.

Walking down a crowded street is quite a different experience if it is dark and we are alone and a friend has just been mugged—then we pay enormous amounts of attention to, and put a lot of mental activity into making conscious decisions about, the other people we see on the street. The trust has broken down, and the rules of civil inattention no longer apply. We panic, too, if someone else breaks the rules: while in the countryside the sight of another human being might be sufficient occasion to provoke a greeting, in the city a stranger who says "hello" as he walks by can be rather disconcerting.

The point with regard to science is that most of the time the public pays it civil inattention: scientists are allowed to go on doing what scientists do, in the background, in the public's peripheral vision if perceived at all. The public do not need to know about science when they trust it. They know it is there, and they accommodate it in their lives, just as they tend not to bump into their fellow pedestrians. But when the trust breaks down—when science seems to have gone wrong somehow—the civil inattention goes too, and suddenly the public are extremely interested to know and understand science.

While we have more and more recourse to experts—be they psychiatrists, counsellors, independent financial advisers, or scientists—we are also more skeptical about them. As a culture, we are coming to realize that experts too have their

own agendas; objectivity and impartiality are very hard to find. We only have to look at experts fighting in controversies to see that there is no single answer any more, and we want to know what is happening and whose agenda is at work. We want to see, as Beck puts it, the whites of science's eyes.

In a world divided up into laypeople and experts, each group needs some point of contact with the other if they are to see the whites of each other's eyes. For the public and science, the mass media are often the only point of contact. (How the mass media work, and how they function as intermediaries between science and the public, is examined in the next chapter.) In situations in which the public feels at risk from science and loses trust—as in the cases discussed in Chapter 7—it is the mass media that provide the forum in which the tensions between the two communities are played out. Then, when the civil inattention has broken down, the public needs to know science and insists that its invisible institutions and practitioners account for themselves in public.

## Chapter 5

# Media Issues in the Public Understanding of Science

**The media are an amorphous, interconnected, mutually dependent bunch.**
Word of mouth, cave paintings, hieroglyphics, tablets of stone, tapestries, manu-
scripts, town criers, books, newspapers, radio, cinema, television: all these have
been, in their own times and contexts, the medium for the masses; they have existed,
in their own times and contexts, alongside personal, private media that they have
never entirely replaced. People talk about last night's TV; radio stations broadcast
books; newspapers review art exhibitions. Media have lifetimes and enjoy vogues:
conversation seems to have survived the challenge of television, as have radio and
cinema, though the public's preferences have changed over time; and it is still tablets
of stone, and not video displays, that sit at the head of our graves. A painted portrait
carries more prestige (or, perhaps, more snob value) than a photograph; yet there is
something more touching about archival footage of a moving, talking human
being—especially one who is now dead, and so appears as if from beyond the grave.
Thus the media can lie to us, without telling any untruths: their potential for literal-
ism is offset by their potential for fiction and by notions of reality as something we
can construct—whether that means staged reconstructions masquerading as news
photography, or the relentless punditry and spin-doctoring of political television.

Media make demands, on both producers and audiences. The media's technology frames, and more than just literally. Television emphasizes pictures and
motion; on radio, sound is everything; cinema is larger than life. Newspapers are a
finite size and come in a whole number of whole pages: if the paper is full and
then one great story comes in late, an event that would have been news is cast into
oblivion to make room for it. Media audiences have to do work: they have to un-

derstand the visual and verbal languages of media communications and absorb their content, if they choose, into their own set of personal and cultural experiences. They are unwittingly schooled in media practices and conventions, so that a viewing experience of a television sitcom is structured according to the assumption that the program will probably last half an hour; a reader looking for sports news will turn to the back of a newspaper, and a radio listener knows that the news tends to be broadcast on the hour.

## UNDERSTANDING SCIENCE-IN-THE-MEDIA

By far the majority of studies of science in the media have been about newspapers. This is not because researchers believe that science in newspapers is the most influential or widespread form of mediated science; nor because newspapers have large readerships: in fact, most studies—in Britain, the United States, and Europe—have been of quality newspapers, which have relatively small readerships (of the order of half a million in Britain, compared to 5 million for a tabloid). Researchers study newspapers, and quality newspapers at that, because that is the most efficient way, in terms of time and money, to study a mass medium. Newspapers are cheap and plentiful; they can be read anywhere and at any pace; they can be cut up, written on, and stored easily; and they are concrete, finite, uncomplicated, there in black and white. The journalistic culture of news has been much studied; and newswriting practices have been codified. Quality newspapers are especially easy to study because they are archived, indexed, consistent, orderly, and clear.

What do studies of science in the media tell us? Professor of journalism Sharon Dunwoody claims that such studies illuminate both the media and the scientific community particularly well because the two professional groups are so dissimilar.[1] Working practices, professional values, type of information—every aspect of the two fields is different, and these differences tell us much about each. But while science-in-the-media is a useful vehicle for understanding the media, few scholars have used it that way: instead, they look at science in the media as a way of understanding science-in-the-media and often end up attributing characteristics to science-in-the-media that are simply characteristics of the media, rather than of the science they see there. This is compounded by the fact that researchers in this area have often come from academic backgrounds such as science, history of science, cultural studies, even law—anything but media studies. But science, as the media scholars are now beginning to demonstrate, is not a special case in the mass media: understanding science-in-the-media has something to do with understanding media science, but mostly it is about understanding the media.

Media studies as a field grew out of studies of political reporting, and other subjects, science among them, have only recently come to the attention of media researchers. Media studies itself is a relatively young field and is awash with models and theories—some conflicting and some complementary. Mass communications scholar Denis McQuail has classified these ideas, and two classes of theory are particularly useful in an assessment of science and the mass media: operational theories, which are about the way in which media and media people work; and normative theories, which are about the way in which the media ought to work.[2] One of the problems of understanding the relationship between scientific and media institutions has been insufficient recognition of the differences between these two types of theory. Journalists, if they offer any theory at all, invariably offer operational theories: they describe how and why they do what they do. (Most often, though, they say they just do what they do.) Media scholars and sociologists usually offer operational theories: they try to understand what is actually happening and why. Scientists, however, and some science communication researchers, offer normative theories: they want to tell the journalists how it should be. Because normative theories are very deeply rooted in cultural and social values, and because these values are very different in the scientific and journalistic communities, the differing perspectives of these two communities have led to considerable friction between them.

The mass media are something about which one can also have an opinion even if all one has done is glance at the morning paper. However, it has been among the tasks of researchers to replace anecdotal evidence about science-in-the-media by more reliable evidence based on tried-and-tested methodologies, as well as to separate "what is" from "what should be." After all, any assessment of the media's potential to fulfill a particular agenda ought surely to be informed by some understanding of how the media actually work.

Research has tended to focus on contemporary circumstances and to ask pragmatic questions reflecting the agenda of the sponsor of the research, such as what percentage of the public would support a particular research program were it advertised in a particular way or what types of book reach which types of reader.[3] The sponsor is, in most cases, some branch or other of the scientific community. The stated aim is often "to improve the public understanding of science," and especially to determine how popularization can produce the particular understanding of science that scientists would like the public to have.[4] This research has thus essentially been conducted to serve the consistent, monolithic aims of the sci-

entific community, and its results have shown similar characteristics. According to communications scholar Christopher Dornan,

> While other areas in media studies have developed via contest or comment, academic discourse on science popularization has been marked by an enduring consensus.[5]

The result of the consensus on how science should be popularized is

> an essentially positivist portrayal of science as a heroic, apolitical, and inherently rational endeavour; . . . "good" science writing should promulgate just this portrayal for popular consumption. In the end, it has served those interests that have found in science a vehicle for the legitimation of the prevailing social order.

High on scientists' agenda (and therefore high on the agenda of science-sponsored science communication research) has been the question of accuracy—that is, the extent to which popularizations faithfully reproduce the facts of science. The scientific community has consistently claimed the right to arbitrate on questions of accuracy in the popular representation of science, and the research literature has highlighted inaccuracy in reporting. This has served to perpetuate an image of popular media as misrepresenting science. Science communication scholars tend not to look at other areas of the media, so while they may be able to measure inaccuracy in science reporting, they are unable to compare that inaccuracy to the inaccuracy of any other field of reporting—we do not know whether science news is more or less accurate than the rest of the news. So finding a context in which to understand science-in-the-media, as opposed to anything-else-in-the-media, is very difficult indeed.

Scientists' insistence on accuracy has produced science journalists who are closely allied with the scientific community, and dependent on it—the journalists rely on the scientists for accurate facts. According to Dornan, this in turn has resulted in science coverage that is more deferential to its subject matter and constituency than would be acceptable in other fields of journalism. Science communication scholar Dorothy Nelkin has argued that the close allegiance between science journalists and the scientific community has produced a false representation of science—one that ignores both the contingency of scientific knowledge and its social and political context.[6] This representation is, however, one with which, as Dornan points out, scientists are content: while it may be "shallow and unrealistic," it fulfills the criteria for good science writing mentioned above. Dornan also points out that not only are science journalists closely allied with the scientific community, but so are many of the academics who study science communication. Thus, while science communication research has engaged critically with the journalistic agencies of science communication, it has only infrequently engaged critically with the role of the scientific community in science communication. Such research has tended to come instead from the sociology of science and the science and technology studies communities, where researchers and their sponsors have a better claim to disinterest.

## MAKING NEWS OUT OF SCIENCE

There are two types of journalists who cover science news: journalists and science journalists. The primary feature of both groups is that they are journalists; and whether they are writing about politics or science or gardening, the same journalistic rules apply. There are certainly some distinct characteristics of the culture of science journalism—science correspondents tend to know each other, go to the same press conferences, and so on—but mostly science journalism is just journalism. Most science in the news is not in science stories and so is not necessarily written by science journalists. But the same factors determine which stories get into the media, irrespective of whether the journalist is a science specialist or not.[7]

Journalists look for stories in many different places. The peer-reviewed scientific literature is a good source, and newspapers often publish stories about work that has already been published in *Nature*, the *New England Journal of Medicine*, *Astrophysical Journal*, or *Cell*. This is an easy (or, perhaps, lazy) strategy for the journalist—the story is reliable, they have all the details in front of them, and they do not even have to leave their desk. It also suits scientists: they do not get pestered by journalists looking for the facts, and, as we saw in Chapter 4, it means that they have also complied with one of the traditional rules of science popularization—they have published in a peer-reviewed journal first and gone public second. This practice is known as the Ingelfinger rule, after Franz Ingelfinger, an editor of the *New England Journal of Medicine*: he decreed that no one shall publish in the *New England Journal of Medicine* if they have already announced their results somewhere else, be that in the local paper or in a science magazine. Most academic journals insist that papers submitted to them be original, that is, not published elsewhere. This is a good example of where the scientific culture and the journalistic culture do not match up: it can take anything from a few months to a few years to get an academic paper into the peer-reviewed literature, but if it is news it has got to be in the newspaper tomorrow. When scientists do go to the newspapers before their peer-reviewed work has been published, they

course—if scientists find the Higgs boson, is anyone else going to give a damn? And for the story to be relevant and consonant, people have to have certain beliefs already, and not many people harbor any impressions or expectations about the search for the Higgs boson. On the other hand, stories about food and health and psychology are meaningful and relevant, because they are about issues that concern everyone, and of which everyone has some understanding. The story of Britain's "mad cow disease" scare over health risks from infected beef is a good example of consonance, because many people were already beginning to think they ought to be eating less red meat. We discuss this story in more detail in Chapter 7.

## Co-option and Composition

If a story can be co-opted—that is, linked to one that is already running—it may have more chance of making it into the news.

Co-option is a very useful news value for science. The Gulf War was a major media event, and on the back of that came many technology stories, pollution stories, and health stories. If someone famous gets sick, it provides opportunities for medical stories about the disease with which this person is afflicted—even if it is one that would not have been newsworthy on its own.

Composition can be considered as a kind of internal co-option. Newspaper readers expect a particular composition: sports at the back, gardening on Tuesdays, lots of politics. A science story is unlikely to be allowed to disrupt this composition of the paper and must fit in whatever spaces are left. If the newspaper has a science section, a story might appear there when it might not have been accepted for a general news page: in a science section, a science story is only competing with other science stories rather than with the latest royal or presidential scandal.

## Frequency, Unexpectedness, and Continuity

Journalists like stories they can anticipate. They can prepare in advance, line up interviews, and reserve a space in the paper. Every Monday there will be the results of the weekend's sports fixtures; every summer a drought; every four years a presidential election. These stories have the news value "frequency." Unfortunately, science does not: the pace of scientific change is nothing if not uneven.

Science may not usually run to schedule, but scientists can take the public by surprise. Science is good for unexpectedness: the public did not anticipate the

1994 comet crash on Jupiter, nor did Americans expect to learn, in 1989, that apples, of all things, could give their children cancer (for more on these news stories, see Chapter 7).

While the rarity value in these events works in their favor in terms of unexpectedness, rare events tend not to be familiar and so score poorly on news values such as meaningfulness and consonance. Moreover, unpredicted events may find journalists unprepared and ill equipped to muster a story before their next deadline.

For busy journalists, another consideration is continuity—for just how long will a story run? An election is a good story because it builds up over a long time to a specific climax: journalists can plan and prepare and build up their store of knowledge and sources over the lifetime of the story. The public becomes involved with the story and familiar with the players and can anticipate the next installment. Some science stories run: mad cow disease is a good example. But the 1996 story about "life" on Mars was over in a few days. That means that journalists who work hard to understand the latest science and to establish contacts with the relevant scientists may only get one story out of their efforts, compared with the dozens that will appear during an election campaign. And science sometimes goes slowly: there may be a next step in a science story, but it might be a long way off; by the time that hard-won scientific knowledge is needed again, the journalist may either have forgotten it or have moved to another posting.

## Competition

Journalists and their editors like to publish stories that their rivals have missed. But scoops are rare in science: because stories crop up erratically, scientific press officers tend not to establish the kind of exclusive relationships with particular journalists that are common in politics or business; and science likes to appear non-aligned, beyond politics, beyond checkbook journalism. Science journalists, too, maintain close links with colleagues: they co-operate between papers and between media— they all know each other and help each other out. Competition does not really come into it—if one journalist has a story, chances are all the others will have it too.

## Unambiguity and Negativity

News needs to have a clear "good news" or "bad news" message. This is why stories like "possible treatment for cancer in 10 years' time" do not rate very highly as news. And the impact of science stories often needs to be explained if this message is to be clear: in what way is this new plastic or new disease good or bad? This means more work for the journalist and more space in the paper. Neither may be possible.

The corollary to "no news is good news" is "most news is bad news." Bad news is certainly more newsworthy than good news. In this regard, science is rather different from, say, crime, where someone has always been done wrong, or sport, where someone always loses. Science is supposed to be for the common good—if the science news is bad, the science must have gone wrong somewhere. Because of the powerful role scientists play as sources of science news, they can keep bad news out of the papers if they so choose. So the science that comes to the attention of journalists tends to be only the good news, which is not always the most newsworthy news.

## Facts, Sources, and Their Reliability

"Facticity" is a good news value for science. Simply from the point of view of news style, news stories require six facts: who, what, where, when, why, and how. Science can usually supply these. A good story needs facts; readers enjoy facts—but they have to be facts that are meaningful, or relevant, or consonant. This is why the fact that 46 percent of women under 26 eat chocolate three times a week is more interesting than the latest value of the Hubble constant. Science scores highly on reliability, because the institutions of science are highly respected. To the surprise of many a scientist, scientists and their institutions are considered highly reliable: all a journalist needs to say is "According to Dr. X of the University of Wherever . . ." and the story is reliable. This is one reason why peer-reviewed journals are such good sources of stories: again, "according to research published this week in *Nature*" makes a reliable story.

## Elitism and Personalization

Some places, and some people, are interesting, and others are not. The small earthquake in Chile was a lot less interesting than an even smaller earthquake close to home. Former colonies, allies, and enemies; neighboring countries; places where readers go on vacation or where their ancestors lived: these are interesting, and what happens there is more newsworthy than similar events in other places. Famous people are interesting, and even minor events in their lives can make the headlines. Science, though, is universal, and who people are, where they come from, or where they did their work is supposed to be irrelevant.

The tendency of the media to personalize stories is a mixed blessing for science. Science is usually done by somebody—there is the "who" already. But according to scientific tradition, it is the "what" and the "how" that are supposed to be more important. Medical stories do sometimes become stories about the patient rather than about the medicine, and the Nobel prizes are news not only because they are big and anticipated, but also because they are about people—after all, the science for which they are awarded is often not very recent. No one will have heard of the scientists before, and they may never hear of them again, but, from the news perspective, at least they are people.

Much scientific research is done by teams: there are not many papers published by only one author, and even those will be just another step in the team effort reflected in the citations that paper contains. Particle physicists tend to work in huge teams, so when they find the Higgs boson, the paper will probably have 40 authors. Journalists need to focus on just one or two people, which can create bad feeling among the ones who do not get into the limelight. Scientific institutions are often very hierarchical, and a university may want to wheel out the head of the department while the journalists want the Ph.D. student who actually did the work. And while the head of the department may have authority and look reliable in the paper, he or she may not have much idea about what actually happened in the lab.

Science clearly has some news value. It has facts and is reliable, and it lends itself on occasion to personalization and co-option. But perhaps the most important lesson from news values is that there are broader forces at work in the selection of stories for news than simply the competing enthusiasms and prejudices of the many interest groups that battle for their share of column inches and seconds of air time. Special pleading for science generally falls on deaf ears, but, as we see in Chaper 6, a science story packed full of news value can land on the front page.

## NEWS LANGUAGE

Nature was all the rage in 18th century France—fashionable as well as popular. Natural history books abounded, and writers took great pains to make their work accessible. One of these books, *Natural History*, became the best-seller of the century. Its editor, the director of the Paris botanical garden, Georges Louis Leclerc, Comte de Buffon, compiled it from the writings of his colleagues but was not satisfied with presenting their work verbatim. As well as giving a scientific account of the natural world, Buffon cast the animals as a variety of characters. Cats, for example, are petty thieves: "pliant and sycophantic like rogues; they have the same dexterity, the same subtlety, the same taste for wickedness; the same tendency to petty pillaging." Some of the scientists whose work appeared in *Natural History* were appalled: anthropomorphism is a literary technique that long predates Buffon, but should a scientist do such a thing in a book that carried such authority? Buffon's colleague Louis Daubenton said that it was simply "a good way of showing off the delights of one's style." On behalf of the lay readers, the influential hostess Suzanne Necker announced: "Animals seem remote from us, and Buffon's art was to bring them closer."

Forms of words are important and contentious in the culture of science. The language of formal scientific publications, and the type of story they tell, are matters of long and precious tradition. They express objectivity, dispassion, and a careful but steady progress toward the truth. An account of what actually happened in the lab, and how the researchers really felt about it, would look rather different. But scientific publications work well within the scientific community, where there is tacit knowledge about the relationship between the words on the page and the work in the lab. These publications have purposes other than the communication of ideas: they are a research team's product and can be used as a measure of their competence and diligence; they define communities and hierarchies in their author bylines and citations; and they are their authors' claim to the small piece of scientific progress they proclaim. Thus, they serve the professional needs of their authors within the scientific community.

But what happens to the words when the information gets outside the scientific community? A useful perspective on this is offered by scholars who have made some pertinent distinctions between three forms of language: the forensic, which states the facts; the deliberative, which makes sense of the facts; and the epideictic, which reacts to the facts and concerns feelings and outcomes. According to rhetorician Jeanne Fahnestock, scientific writing prioritizes the forensic over the deliberative over the epideictic, and popular writing—what Fahnestock calls "accommodations"—prioritizes the epideictic over the deliberative over the forensic.[11] Scientific papers, for example, are largely forensic: they are about the facts—what was done and what data were produced. Because the author is an expert and is writing for people with the same expertise, much goes unsaid: "this would make a great new drug" or "at this point we all got really excited" will be inferred by expert readers. Significance is understood. But, suggests Fahnestock:

> Accommodations of scientific reports, on the other hand, are not primarily forensic. With a significant change in rhetorical situation comes a change in genre, and instead of simply reporting the facts for a different audience, scientific accommodations are overwhelmingly epideictic; their main purpose is to celebrate rather than validate. And furthermore, they must usually be explicit in their claims about the value of the scientific discoveries they pass along. They cannot rely on the audience to recognize the significance of information. Thus the work of epideictic rhetoric in science journalism requires the adjustment of new information to the audiences' already held values and assumptions.

News langauge is immediate, positive, and active, unlike the measured, passive prose of science.[12] In a news story the important information is right at the beginning of the article: the first sentence tells the whole story and includes as many of the key facts as possible; details, technicalities, and qualifications will be toward the end of the article, if they are there at all. Science is supposed to be generalized and generalizable—it deals with classes of things and repeatable events; but news is specific, so the words used must be specific, rather than general: "gun" rather than "weapon"; "pig" rather than "mammal." This makes the news story more concrete and easier to relate to and to visualize. It may also skew the emphasis of the original science story; and, by making it more accessible, it may provoke a more emotional response in the reader. To be concise, journalists may remove the qualifications in scientific langauge—"this is very good, but only under these conditions" becomes "this is very good." This has the effect of exaggerating—and exaggerations have greater epideictic appeal. It also makes information look more certain than the scientists might think it actually is.

To be relevant and meaningful, news reports often emphasize the potential applications and outcomes of scientific results, rather than the process by which they were developed. Emphasizing applications again makes the information seem more certain—already the results have some use in the real world, so they must be right; and the results make sense to us, because we connect them with a real-world problem—a problem which may provoke an emotional response. Journalists, who may be non-experts in science and have no professional stake in it, can afford to speculate—and anyway the paper will soon be thrown away and forgotten.

This shift toward the epideictic in popularizations of science—and the sensationalism and gee-whizzery that may result—is, then, not a punishment inflicted on science stories by disdainful or malicious journalists. It arises in science journalism because of the rhetorical conventions of popularization—conventions that apply to all journalism, whether about science or not. Science journalism is primarily journalism.

## WHAT SCIENCE, AND HOW MUCH OF IT, IS THERE IN THE MEDIA?

It is easy enough to draw conclusions about science in the media from one's own experience of reading and viewing; but arriving at a broader picture as a result of rigorous research is more of a challenge. That there is science in the mass media is undeniable, but how much is there, and how is that science represented? These questions are answered to some extent by content analysis, which allows researchers to draw quantitative conclusions (how much or how many) about qualitative features (subject matter or "tone"). For practical reasons, most content analysis has been of text. In studies of science in the mass media, most content analyses have been of newspapers.

In order to do a content analysis, a researcher will develop a "coding frame" consisting of questions, possible answers to which are represented by a code. Each question is about a different quality or quantity that the researcher wants to gauge. Some of the questions will be straightforward: how long is the article, or where does it appear within the paper? Others will require a judgment on the part of the coder: is the article positive or negative in tone, or who is the main agent of the story? Content analyses try to look at articles from several different perspectives: each article reflects decisions made by, for example, the journalist, who decided that the story was worth writing and then wrote it in a certain way; by the editor, who decided to run the story and who put it in a particular place in the paper; by the copy-editor, who decided how long each article would be and what the headline would be; by the photographer, who took a photograph with a particular composition, and by the picture editor, who decided whether or not the photograph was worth including. Some of these decisions reflect newspaper practices: news stories have a particular format; pictures will be spread evenly through the paper. Other decisions reflect cultural or professional values: is this a good story, and what is important about it? Some aspects of a story will be determined by the content and origin of the story itself; others are determined by the context, in terms both of the news on a particular day and of the society in which the news is published. Content analysis tries to come to grips with all these variables.[13]

As a result, researchers face a dilemma: the simpler the coding frame (fewer, easier questions), the more manageable the data—but the less interesting; the more codes (more questions), the more interesting the result, but the trickier the coding and the more complicated the analysis of the overall data. Typically, though, a coding frame might include on the order of 50 codes. Researchers are always striving for better coding frames, and each frame represents a research group's particular agenda. This has the unfortunate consequence that every time a frame is "improved," its data are no longer comparable with data from the original frame, which makes comparisons between studies difficult.

One big question which faces this type of research is what counts as science coverage. Some science stories are easy to spot: they are on the science page and are written by a science correspondent. Others are ambiguous: is an article on organic pest control on the gardening page a science story? Or an article about a car-crash victim who just happens to be a biochemist? And what about an article about how 42 percent of the adult population prefer coffee to tea? The solution lies not so much in which answer one chooses as in how one assesses its reliability. Within an agreed definition of a science story, coders choose which stories they think count, and the consistency with which they do this, both as individuals on different days and between coders, is measured and reported. But, again, research decisions affect the answer: the narrower the definition of science, the less science is found in the media.

The Science Museum Media Monitor is an ambitious content analysis project on science in British newspapers. It comprises 6000 articles sampled from quality and tabloid newspapers from 1946 to 1990 and coded on 70 variables.[14] It used a broad definition of science, which included social sciences such as economic research and opinion polling, and identified articles by their use of jargon, citations, and technical illustrations, as well as by explicit mention of science, technology, medicine, or the environment. There were some specified exclusions: among them were astrology, weather reports, maps, and voter research. The study, which was led by social psychologist Martin Bauer, took a team of five people thousands of working hours.

The Media Monitor yielded some interesting results. In the quality press, coverage increased massively in the period 1946–1960, then decreased until 1974, and then increased again so that in 1990 levels were comparable with those of 1960. In the popular press, a similar post-war increase leveled out from 1962 to 1978, and coverage then declined during the 1980s. Over the whole period, science has occupied about 5 percent of the total space in newspapers, with more science news in quality papers than in the popular press.

The tone of coverage has changed over the years. Until the mid-1960s, the tone was generally positive and celebratory; since then, the tone has changed to negative and critical. However, at the very last point in the study (1990), the tone was again tending toward the more positive. Until the mid-1960s, coverage stressed the benefits of science. Since then, there has been a slight decline in the number of stories that describe the benefits of science, a sharp increase in the number of stories that look at the risks of science, and a considerable increase in the number of stories that weigh the risks against the benefits. Overall, risk has become a major feature of newspaper science. Controversies, however, feature in only a quarter of science news articles and were most prominent in the late 1940s and early 1980s.

During the period studied, there has been a shift in coverage from an early emphasis on the physical sciences toward the social and biomedical sciences (although the popular press has tended to ignore the social sciences). In terms of the subjects covered, the Media Monitor has revealed a repeating pattern of coverage of technologies that have a particular social prominence. In the 1950s, nuclear technology was uppermost; this was then replaced by space and astronomy, which peaked in the early 1960s but continued to attract attention into the 1970s. Information technology peaked in the mid-1980s and was overtaken by biotechnology and genetic engineering toward the end of the 1980s.

It has been claimed that science gets poor treatment by the press because it is impossible to explain it properly in a short article. The answer to this would be the feature article, which tends to be longer and so has more room for explanations and background. The Media Monitor found that, over time, an increasing proportion of science in the newspapers has been covered in features rather than in news.

A content analysis of science in U.S. daily newspapers has been published by communications researcher Marianne Pellechia, whose particular aim was to test whether charges by scientists that journalistic accounts of science omitted important information, such as qualifying statements, background, and methodological details, were justified.[15] Looking at three four-year periods from the last 30 years, Pellechia found that while science news represented only a small percentage of news coverage, the amount was steadily growing, but the character of the coverage remained constant over that time. Pellechia used a narrower definition of science than the Media Monitor study: science news should have substantial subject matter concerning the results and interpretation of empirical research in science, technology, engineering, or medicine. Local stories, syndicated stories, and editorials were not included. The coding frame was based on several others from previous studies; so while the other frames had all been road-tested and their reliability determined, Pellechia's was still different from each of the frames used before. The articles were sampled from the broadsheet newspapers the *New York Times*, *Chicago Tribune*, and *Washington Post*, which Pellechia thought were least likely to show the kind of omissions usually lamented by scientists.

Compared to the 5 percent science content found in British newspapers, Pellechia found a figure ranging from 0.5 to 2 percent (the different definitions of science may account for this difference). Again, there was a big emphasis on medical stories—three-quarters of all stories were medical or related to health. As in Britain, science articles in the United States are getting longer. There was a slight increase (from a very low baseline) in the number of articles that included a second opinion, either positive or negative, from a scientist other than the main subject of the story. Otherwise, the stories were very consistent with respect to the aspects coded: omissions included identification of researchers and their institutions, citation of prior research, qualifying statements, and information about methodology—all aspects that science communication researchers and scientists think should be included. Pellechia does not speculate about why journalists seem to think that this sort of information has no place in their newspapers; she does, however, remark that despite the many calls to remedy these omissions, nobody really has any concrete evidence that including the missing information would make any difference at all to the readers' understanding of science.

Content analysis is in many ways a very useful technique, and it provides reliable information against which to judge the claims of interest groups and complainants. A study on the scale of the Media Monitor also indicates that broader forces are at work shaping science in media—social trends and professional practices loom large in the data. But what content analysis of science in media does not do is situate the media products in the contexts either of their production or of their reception.[16] For example, was the critical tone of science reporting in the 1970s a reflection of a wider social turn against science; or was it that science journalists were disenchanted with science; or was journalism as a profession simply becoming more critical across the board? And when the public read these critical articles, did they turn against science or mentally rush to its defense? However, given the many forces that shape science in the media, and the many contexts in which it is received and understood, it is perhaps remarkable that we can know anything about it with any certainty; and the enormous challenge that this research presents is perhaps the reason so many people still rely on their own impressions when they express their opinions of science in the media.

## SCIENCE ON TELEVISION

Great works of natural history dominated popular science right up until the end of the 19th century. Some of these are classics of the genre: Buffon's *Natural History* was the standard text for scholars, rambling parsons, and lady salonnieres for over a century, and Gilbert White's *Natural History of Selborne* is still in print after 200 years. One reason why many natural history books were so popular was that they were full of pictures. The inadequacy of words could be compensated for by artists who often became famous in their own right—John James Audubon's birds and the flowers of Jacques Le Moyne de Morgues thrive as works of art long after their value as a scientific record has gone. Yet just when increased opportunities for travel opened up the world and its creatures to the eyes of the curious, another revolution swept into our homes and saved us the bother of ever going anywhere. Television rounded up the natural world and placed it, safe, clean, and odor-free, under our very noses. The contemporary artist of natural history has abandoned brush and pen and now operates a TV camera.

Natural history has been the triumph of television science, but, as we saw in Chapter 2, a broad spectrum of the sciences have found their way, in many different guises, onto television screens around the world. Documentaries are a long-standing, successful format for science: the one-hour, single-subject film seems to cope well with the demands of entertainment and explanation that audiences need from their TV science, and they sell internationally—"Horizon" and "NOVA" are successful around the world. Current affairs programs often look at the social and political circumstances of science, technology, and medicine. Drama series about science issues have been successful in terms of audience numbers, and science plays a key role in many other dramas, especially crime stories and medical soaps. Science fiction series have boldly gone where no series have gone before to bring the language and scenery of science to new audiences. Magazine programs trade in topicality and gadgets; and science has become an important component of TV news. Children are an important audience for TV science, and programs less bashful about their didactic intent have been popular with young people. Even cartoons often have a science theme.

How, then, does all this science shape up? Studies indicate that what interests people in newspapers also interests people on television, so medicine and nature score highly in terms of amount of coverage.[17] Technology, social sciences, and environmental issues are also popular subjects for TV science. Nature has the added benefit of cute subjects and striking landscapes—TV assets the particle physicists are hard-pressed to match. Television requires pictures all the time, and those pictures are required to change every now and then. It requires sound most of the time, and a degree of variety in that sound. But the pictures and sounds need to be meaningful and relevant: very little of the science we experience in school looks and sounds like something you would want to sit with in your living room. And a lot of science does not look like anything at all: gravity, the greenhouse effect, oxidative phosphorylation—visually, they do not have a lot to recommend them. So it is really no surprise that there is not much particle physics or biochemistry on television: they do not look like anything. Nature, on the other hand, is full of pretty pictures; the environment has great scenery; technology is on a human scale and does something we want it to do; and the social sciences are about us.

What scientists actually do is not very televisual: science happens inside people's heads, or in small gestures—a modification to a computer program, another gram of reagent. Even the big things happen slowly. An extraordinary "Horizon" about the discovery of the W-boson at CERN (European Center for Nuclear Research) was a film of a bunch of guys in shirtsleeves who worked in a place full of steel and pipes and wires, and the big climax was that they found this tiny thing that they could not even see, let alone show to the cameras. But the blip on the graph was an important event, the producers said, and it deserved to be recorded—even though it made lousy television.

But for most science, producers stick with the conventions of television. So how do they react to the lack of televisuality of a lot of science? There are two responses. The first is to concentrate on those parts of science that are easily visualized. Volcanoes, explosions, and lightning make good TV because they move and they look and sound like something. Astronomy is not bad: it does not do much, as far as anyone can see, but it is pretty, and there is plenty of scope for graphics. Graphics are a fiction of sorts, but they are the sort of fiction that scientists themselves use, as a glimpse at any textbook will demonstrate; and the computer graphics that television uses to render science visible are also the means of much scientific research, be it into comets or viruses. The stars of David Attenborough's series "The Secret Life of Plants" did not move—until he speeded up the film. The fiction of plants blooming before our eyes was a technique that television used to make science look like something, and which also communicated information about life cycles and changing seasons by letting us watch them happen in a few minutes.

What do all these transformations do to science on television? The very big and the very small are very hard to understand, even if we can see them. So small things become bigger and big things become smaller—whatever it takes to fit them onto the screen, and into our cognitive framework. Fast things get slowed down and slow things get speeded up—science needs to be slow enough for us to see it but fast enough to tell a story in a mere hour. Blurred pictures become crisp graphics; atoms are colored and held together by sticks; galaxies form in seconds. Many of these transformations happen wherever one deals with science: even in the most intimate relationship between mind and nature, the scientist is conceiving of science in terms of models and images, transforming nature into something we can grasp. And what do scientists do on television? They talk to the camera or to the interviewer, and they go jogging, have lunch, get in and out of their cars, walk up and down corridors—but they do not actually do science. Science rarely happens on television, though somewhere off-camera a problem will be posed, solved, and celebrated within the hour.[18]

Television carries no flag for science: it has no responsibility to represent science in the way scientists or teachers or politicians would like it represented. But what about the public? Are they being conned by science on television? What about all the invisibilities of science: do the the public know that there is something going on that they do not see? Are scientists simply people who sit in front of bookcases and talk to journalists? Is that how science is done?

Television tells stories about science, and the stories it tells are in many ways the same as all stories ever told. Stories need characters—heroes and, possibly, even villains; they need plots, denouements, beginnings, middles, and ends. According to media studies professor Roger Silverstone, there is always a tension between the storytelling aspect—in his words, the "mythic" aspect—of television science and the need for a realistic representation of the scientific processes at work—Silverstone's "mimetic" strand of the program.[19] A good science story in television might start with a problem, set out mimetically in terms of "what were, and what killed, the dinosaurs?", and then the mythological hero will step into the frame and start thinking about how the problem might be solved. With possibly one or two colleagues—why burden the audience with any more?—the hero will take a series of cleanly executed steps toward the goal; and, lo and behold, in precisely an hour the problem will be solved. Very rare indeed are programs about, for example, why it is so difficult to develop a vaccine for HIV or why scientists still do not know what dark matter is—as rare, perhaps, as police dramas in which the bank robber evades capture and lives a life of luxury in the sun. If the overall social message of police dramas is that crime does not pay, the overriding social message of science on television is that scientists always solve the problem, even though how they do so must remain invisible to the public. This is a message that, according to the arguments presented in Chapter 4, can do little to further the public understanding of how science really works.

# THE FASCINATION WITH UNORTHODOX SCIENCE

Journalistic practices and news values can account to some extent for the attention alternative and fringe science attracts from the press. Amateur and unorthodox scientists routinely publish in the public domain, and even conventional scientists who have been unsuccessful in publishing their work in the peer-reviewed literature may look to popular media for an outlet for their work. For some subscribers to the dominant model of popularization (see Chapter 4)—those for whom popular science is bad science—the public domain is the rightful place for fringe science. But other scientists worry about the effect unorthodox science might have on the public's understanding or (or, perhaps, appreciation of) science (see Chapter 3), and some scientists even accuse journalists of betraying "real" science when they report unorthodox science.

If science that appears in the public domain has not first been published in the peer-reviewed literature, then the scientist has broken with an important professional tradition. Of course, if one has been unsuccessful in publishing in the peer-reviewed literature, and one wants to follow the rules, one should give up any attempt to publish anywhere. But fringe scientists have a message to get across, just like their more orthodox colleagues. In 1950, Immanuel Velikovsky proposed an unorthodox cosmology that would have rewritten history, astronomy, and biology. His book, *Worlds in Collision*, caused an enormous outcry among scientists, and it became an international best-seller. It was published by Macmillan, and scientists in the United States organized a boycott of Macmillan textbooks. To protect its market, Macmillan sold the rights to *Worlds in Collision* to Doubleday, a company that did not publish many textbooks and so was relatively immune to academic pressure. So the controversy rumbled on, and in 1956 *Scientific American* published a satirical article about the book. When Velikovsky asked for right of reply, the editor of *Scientific American* said that Velikovsky's views had done incalculable harm to the public understanding of science, and since they had already reached far more people than they possibly could in a scientific magazine, he did not feel he could help.[20] This exchange reveals two tacit understandings: that scientists would already have seen Velikovsky's views since the book was in the public domain; and that for Velikovsky, being published in a scientific magazine still mattered in an important way, for which no amount of sales royalties or public attention could ever compensate.

Several studies have been carried out on media coverage of an unorthodox earthquake prediction in the United States. One such study, by mass media scholar

Conrad Smith, reports how a prediction by the U.S. Geological Survey (USGS) of an earthquake that did happen was reported only 8 times, while an "unscientific" prediction by Iben Browning, a self-trained climatologist with no formal qualifications, of a different earthquake that did not happen was reported 68 times.[21] Smith suggests that several factors were at work here. Journalists make news by personalizing, simplifying, and symbolizing. They tell stories about events rather than about issues. They interview people who are available to them (who may not be the most authoritative). Non-specialist reporters may not know who is an expert and who is not. News is about a person, not about a group of people; about conflict, not consensus; about the facts that tell the story rather than the facts that explain it. And news is about drama.

Iben Browning's prediction was much more specific that the USGS's: it specified a four-day period while the USGS prediction suggested a 30-year period. Browning suggested an actual town, New Madrid, Missouri; the USGS suggested the San Francisco Bay area. Browning had apparent authority because he was a Ph.D. (in zoology); and he based his prediction on tidal forces, which looked scientific, even though, according to orthodox science, there is no connection between tides and earthquakes. Browning's prediction, being more precise, was more dramatic; and once the drama had worn off, the newspapers reported public fears. They had no reason to do this with the USGS prediction, which was imprecise, undramatic, and, therefore, scared no-one. Five months after Browning's prediction, the USGS took the unusual step of publishing a statement which said that Browning's prediction was meaningless. Before the statement, 11 percent of newspaper stories questioned Browning's scientific credentials; after the statement, 31 percent of stories did—still less than a third. After the USGS statement was published, and journalists had some names of orthodox scientists in front of them, they did contact them. Interestingly, whether or not the scientists interviewed had good credentials made no difference to the accuracy of the stories. Nor did non-specialist reporters quote less well qualified scientists than specialist reporters: they knew not to rely on their own limited scientific expertise and did a good job of checking for scientific credentials.

Browning's prediction of the four-day period December 1–5 was simplified to the very specific "December 3," which made it sensational; the USGS prediction of 30 years remained 30 years. After the USGS statement had been reported (over a period of two days), the newspapers continued to report the buildup to the predicted December 3 earthquake but did not refer again to the USGS report. The newspapers reported the possible consequences of the earthquake by asking people how it would affect them. This brought the earthquake into the real world and made it seem more certain. Reporters talked to local people on the streets, who were more available as quotable sources than scientists in institutions. Communications scholar James Dearing suggests that reporting the views of people in the street with whom readers can identify can give unorthodox science credibility among non-scientists.[22]

A telephone survey conducted at the time showed that abut 50 percent of people thought that Browning's earthquake would happen, so the news reporting, even after the USGS statement, did not manage to combat the "scientific error." Since the journalists relied entirely on the USGS report for their criticisms of Browning, if the report had not been issued, they might not have criticized at all. This suggests that journalists are manipulable by scientists, reinforcing the dominant model which puts scientists at the top of the information hierarchy and confirming journalists' dependence on scientists. But Dearing suggests that pitting an unorthodox scientist against the scientific establishment (in this case, the USGS) in a news article would have excited public sympathy for the lone outsider. Dearing also suggests that the public are more likely to believe a maverick because the public make judgments on a broader range of criteria than scientists—including fear, superstition, and gut reaction. Thus, for many people, science is but one among many factors that inform their decisions and reactions; and so the fact that a prediction is declared unscientific is unlikely to make much difference to its credibility.

Dearing's study of maverick science concluded that stories about mavericks tended to be more subjective than objective; that is, they took the mavericks on their own terms and did not criticize. Such stories tended to support the maverick, but in 70 percent of the stories in which a second expert was quoted, the expert was critical. Dearing suggests that this criticism only serves to give credibility to mavericks: since we are used to political reporting, when we see two experts on opposing sides arguing, we think they are on level ground. In the case of unorthodox science, what we do not see is that one scientist represents the majority opinion of the scientific community, and the other represents only himself. While 66 percent of journalists thought that mavericks are useful for scientific progress, 82 percent also thought that journalists themselves should be more critical of them. Dearing found that while journalists were giving mavericks a chance to communicate their ideas on their own terms, most of the journalists did not actually believe the ideas they were communicating. This is perhaps not so surprising, given the strong traditions of political reporting in the culture of journalism, where at least half the time journalists must be reporting ideas they disagree with. Science, however, is not a multi-party democracy.

# WHAT EFFECT DOES MEDIA SCIENCE HAVE ON THE PUBLIC?

Despite the long-running battles between scientists and the media, and the often-voiced pleas for more science in the media, no one really knows much about the impact of mass media science. How much of it makes any impression at all on readers and listeners? In what sense, if any, do the media make a difference? There are two angles from which common sense gives us a useful perspective. The first is in the numbers. Fewer than two million people read the *New York Times*, corresponding to less than one percent of the U.S. population. Four million people read the British tabloid the *Sun*: still only seven percent of the British public. There are some multiplying effects though: because news values apply universally, much the same stories will be featured on the evening news as appeared in the evening papers—you may not be the one in 100 who reads a particular newspaper, but you might be the one in 30 who watches the news bulletin. People talk to each other about what they read in the newspapers or what they see on television. Even so, in the term "mass media," "mass" rarely means "everybody."

The second piece of common sense is about the audience experience. Anyone who reads a newspaper will know that they probably read only about a tenth of the articles, if that—indeed, newspapers are designed to facilitate skipping, skimming, and selective reading. Readers have a tacit sense of the news rules: if the Pope has died or if war has broken out, it will not be necessary to scour page 15 to find the announcement: a glance at the front page will suffice. Of the millions of people who read tabloid newspapers, maybe only one in 10 will read a particular article, and maybe only one in 10 of those will think it important or memorable, and maybe only one in 10 of those will tell someone else about it. Memory is a big problem: news has quite a short half-life in most people's minds, and even if you did watch a news bulletin last night, chances are you will not now be able to remember what was on it.

On the other hand, we are as individuals very poor judges of the effect our experiences have on us; and we all know that some stories make a big difference, albeit briefly. Media coverage of suspect foods keeps people away from those counters at the supermarket; doctors are inundated with people who think they have a rare disease every time a cluster of cases is reported. So: what does research have to say about the impact of the media on what people think and know and do?

There have been many different schools of effects research in media studies, and the ongoing debates about the influence on viewers of media violence and the impact on voters of media politics indicate that such research has some way to go before it can provide us with any conclusive insights into the complicated interactions between people and the media.[23] Some research looks at the phenomenon of media effects as a simple matter of transmission: the media emit a signal, and the audience receives it—there is no sense that people might be part of a culture in which they interpret and construct meanings and accommodate them in a particular form within their own experience. Another claim is that media effects are actually very small and limited but have been overestimated by people who have a vested interest in constructing media power as a cultural given. Interest groups include not only media practitioners and people working in such fields as advertising and public relations who rely on the assumption of media effect for their livelihood and career, but also media effects researchers, who would be biting the hand that feeds them if they were to conclude that they are studying a non-phenomenon. Less cynically, perhaps, research conventions demand results, rather than non-results: no one publishes a paper that identifies a non-phenomenon. So media effects research tends to identify small impacts and to conclude with a plea for more research. But the question is: do such studies indicate that media effects are small because the effects are indeed small, or because we do not yet have an efficient method for measuring them?

The difficulties of media effects research are legion. One problem is in deciding how much of an effect is actually due to the medium: if a politician is persuasive on television, does the politician take the credit, or the television? Where is the cause of the media effect? And then there are the audiences and readerships: people are very tricky subjects to study. A message may get through, but is receipt of a message an effect? If the message is "eat less fat," is it enough that the message is received? Surely, the desired effect is that less fat gets eaten—behavior must change if there is to be an effect. The message may be competing with other media messages that are aimed at producing the opposite behavior: advertisements for butter and chocolate, if they are persuasive at all, may be just as persuasive as the public health announcements about eating less fat.

When do you test if a message has been received: after two hours? It may soon be forgotten. After two weeks? How long does a message need to reside in the mind if it is to change behavior? "Wear white at night" is applicable every day, but "nuclear weapons are necessary" needs to be remembered until the next election. And if behavior does change, how does one establish a causal connection with the media message, when there are all sorts of other messages, beliefs, and attitudes that inform and guide human behavior? These problems confront all media effects researchers.

A study by communication researcher Fiona Chew and her colleagues illustrates some of the challenges researchers face when trying to assess the impact of media science.[24] The idea was to see whether a single television program giving health advice "made a difference." First, the team telephoned 1000 people to tell them the program, "Eat Smart," was going to be broadcast and to ask them some questions to see what they knew about nutrition and health. The researchers then called the same people again after the program had been shown, and the 400 or so who had watched the program were asked the same questions again to see if they got more answers right. Then these people were called again six months later to see what they remembered.

The researchers faced many difficulties in gathering and interpreting their data. First, there was no testing of people who had no idea that the program was on and did not watch it. The advantage of doing this would have been that these people would have been subject to all the same other effects—all the influences sloshing around in everyday life—as the people who did watch the program, except for the significant fact that they would not have watched the program. So the effect of a single media product—that particular program among all the others—would have been easier to pinpoint. The researchers were unable to do this because it would have doubled the cost of the study.

Second, although the people who watched the program were not told beforehand that they would be quizzed afterward, they had been quizzed before. Those who had found the questions difficult might have been alerted to their own ignorance and watched the program in order to educate themselves. Or they might have chosen not to watch the program because they felt so ignorant that they decided they would not understand it. Those who found the questions easy might have watched the program because it was about something they were familiar with—or not watched it because they thought they knew it all already. Whatever the case, the people who were telephoned would have associated the program with having or not having knowledge about health and would have watched it in that frame of mind. This is not the way people usually watch television.

So the result that the people who watched the program knew more afterward is not too surprising; the viewers were expecting a learning experience, and that is how they viewed the program. The fact that they remembered some of what they had learned for quite some time afterward is also not surprising, because, after all, they had already, by that time, been quizzed twice by the person on the telephone who wanted them to watch the television program. The people who chose to watch the program might have been more interested in health issues, and so would have been more receptive to the information in the program; and they might have found it easier to retain and accommodate the information than someone who is not interested in health—that is, if any such person watched the program in the first place. Nor is it particularly surprising that more people mentioned television as a source of health information after they had watched the program. After all, they had just used the television as a source of health information, as had been suggested to them by the voice on the telephone.

More people after the experiment than before used their newspaper as a source of health information. This suggests that some people, who may have been newspaper readers anyway, had been sensitized to health information by this experiment and were then more likely to notice health news and information in the newspaper. This result is consonant with the results of other scholars who suggest that while the media may not necessarily tell people what to think, they do tell people what to think about.[25] But again, was this effect, in the case of "Eat Smart," the result of the television program itself, or of the interaction with the researchers on the telephone? So the program increased knowledge slightly, but did it make any difference to behavior, and hence to health? The study by Chew and colleagues makes no claims on this score: the practical problems of gathering such information are easy to imagine.

A review by communications researcher David Gauntlett of effects research concluded that we might reasonably expect mass media public health information campaigns—stop smoking, eat better—to be effective on the level of a few percent here and there; maybe three percent stop smoking, and two percent eat more vegetables.[26] This is not a big effect, except in terms of individuals: a public information media campaign in Sydney to get people to stop smoking produced a two percent increase in the quitting rate, which amounted to about 80,000 extra people who gave up smoking. That means that nearly four million people did not stop smoking, but the fact that 80,000 did will make some difference to people's lives and the economy and health resources.

Campaigns that are particularly aimed at having some specific effect seem to work much better, Gauntlett reports, when there is also some connected personal contact—when the viewer, the recipient of the message, actually does something to find out more or to change their behavior, and does it with the help of another human being, such as a doctor or a friend. Mostly, though, direct information campaigns are not efficient in terms of changing behavior; the suspicion is, among media researchers, that drama campaigns—hidden messages in stories (parables, fables, soaps)—work much better. What media campaigns may do is alert people to issues and sensitize them, and they may give people vocabulary that enables them to access other sources of information or opinion. And, as Gauntlett remarks, at least if the media are not much good at doing good, chances are they are not doing much harm either.

Research on the influence of media presentations of science that are not specifically focused on presenting a particular information message is even more challenging. If there is no specific message—if the communication conveys impressions or creates attitudes, rather than imparts knowledge—it is very difficult to test for the impact on the recipient. For example, many scientists cite television as something that got them involved in science. But what it was about the programs they watched that enthused them in this way tends to be rather elusive; we also know that millions of people watched the same programs and never became scientists.

We communicate in caricatures and stereotypes: in police dramas, the bad guys are never as good-looking as the cops. When media scientists wear a white coat, the coat is the label or badge that makes the story easier to follow. (Perhaps scientists should be thankful: white was always the color of the good guy's hat in the cowboy movies.) Yet, despite the "mad boffin" media images, survey researchers repeatedly report that the public thinks that scientists are ordinary people.[27] Ask people to draw a scientist, and they will draw a wild-haired man or a bespectacled bald man—it is always a man—perhaps with a white coat and bow tie. But they draw that not because that is necessarily what they think scientists look like, but because that is what they think other people will recognize as a scientist: they opt in to the caricature for the purposes of communication, just like the mass media. The practice of science may be absent from documentaries, but it is exhibited in the problem solving in science fiction shows such as "Star Trek" or in the tests and procedures of medical soaps such as "ER." These present a vocabulary and rhetoric of scientific life. But how they impinge upon people's impressions gathered from real-life sources, we do not know.

Does media science make a difference to how people think about science? Possibly, but we do not know. How can we separate the influences of a particular medium from those of other media and from other sources of impressions and influences? We cannot. Do media give people information? Yes, sometimes, to a certain extent. Does that information have any effect? Sometimes, for some people: but when, and how much, and for whom, we do not know.

For the scientist who wants to be a source of science news, forewarned is forearmed: collaborating with journalists and adapting to journalistic conventions may give scientists more, rather than less, control over the emphasis and tone of the resulting story. But the last word will always go to the journalists, because science journalism is much more about journalism than it is about science. For the public, understanding the implications of journalistic conventions for the representation of science in the news may offer some insight into the science behind the headlines.

**Chapter 6**

# Case Studies in Public Science

**So far in this book we have examined the models, concepts, and politics of the** public understanding of science. In this chapter, we look at some episodes of public science to see how these ideas work in the real world of science in public.

The mass media most often provide the forum in which science and the public meet. However, as we discussed in Chapter 5, the media, in reflecting scientific developments and transmitting them to the public, also help to shape the course of discussions within the scientific community; and the coincidence of scientific tensions and media agendas can amplify stories by connecting them to larger concerns. The reporting in 1992 of results from the NASA satellite CoBE, whose "eyes" could see back to the afterglow of the big bang, is a case in point. Here we examine ways in which the media and the scientific community interpreted this story.

Sometimes, the media seem to be setting the agenda for individual scientists if not for science as a whole. In a remarkable three weeks in 1919, the press thrust Albert Einstein from obscurity into the limelight that he was destined never to leave. We examine the events and press coverage of the eclipse expedition to test Einstein's theory that brought him to public attention and ask: Was relativity "proved" by the London *Times*?

In Chapter 4 we outlined the formulations due to John Durant of the idea of scientific literacy. He discusses three of the definitions that are used by various players in the field of public science, namely, "knowing a lot of science," "knowing how science works," and "knowing how science really works." The real-life situation of an oil pollution accident offers a useful test of such models, in an area where science merges with technology. In this chapter we examine the environmental science reported after the tanker *Sea Empress* ran aground off the Welsh coast in 1996 and ask what would constitute scientific literacy for someone wanting to understand the accident and its aftermath.

We begin, however, with some examples from chemistry, which has historically produced some of the paradigms of good science communication but which, today, is something of a poor relation in popular science. Chemistry was arguably just as crucial as physics in the "scientific revolution" of the 17th century; and it has shaped our modern vision of the universe. But nowadays you would hardly think so from looking at the popular-science shelves of any bookstore or running through the catalog of popular-science programs available from public service television. To be sure, a few stalwart chemists fight their corner; but their works barely cover a postage stamp compared with the acres of bookshelves devoted to popular accounts of quantum physics and relativistic cosmology which, in turn, pale into insignificance alongside the massed ranks of modern genetics and natural history potboilers for public consumption. But it was certainly not always thus, as the following story shows.

## CHEMISTRY, CANDLES, AND CARBON

### The Bright Light of Faraday's Candle

For nearly two centuries, the Royal Institution in London has been the venue for public lectures on science. So popular were the talks given there by Sir Humphry Davy in the early 1800s that the police had to make the road outside one-way to horse-drawn carriages—it was the first traffic management scheme in England. In 1826, Michael Faraday, who succeeded Davy as professor of chemistry at the Royal Institution, started a series of Friday evening discourses, which were popular with mainly middle-class audiences keen to keep up with the latest scientific thinking. Lectures offered as a Christmas treat for schoolchildren began not long afterward; these are still given by leading scientists, and broadcast on national television, during the Christmas school recess.

Faraday was one of the leading scientific thinkers of the mid-1800s, but he was also ranked highly among those who brought science to a wide audience. A popular technique of the time was to take an everyday object and show how science gave important insights into its workings. Thomas Henry Huxley gave his audience of working men and their families an entrée into the worlds of geology and palaeontology through a piece of chalk,[1] but Faraday used the humble candle: "There is no better, there is no more open door by which you can enter into the study of natural philosophy, than by considering the physical phenomena of a candle," he declared.[2] Faraday's six lectures on "The Chemical History of a Candle" were first delivered at Christmas in 1848 to an audience that included the young Prince of Wales and Queen Victoria's beloved husband Albert.[3]

Faraday led his audience from the manufacture of candles and the nature of their flames to atmospheric chemistry and respiration. His talent lay in getting his audience to think more deeply about things they usually took for granted. The wax candle was an enigma that he was about to explain; but once he had explained it, anyone could do it—or at least they could while the warm glow of the Victorian evening lingered. Faraday posed the question:

> You have here a solid substance with no vessel to contain it; and how is it that this solid substance can get up to the place where the flame is? How is it that this solid gets there, it not being a fluid? Or, that when it is made a fluid, then how is it that it keeps together? This is a wonderful thing about a candle.

Faraday then took a light to his candle, an act that his audience must have carried out several times a day during the winter months. But now he wanted them to think about it:

> You see, then, in the first instance that a beautiful cup is formed. As the air comes to the candle it moves upwards by the force of the current which the heat of the candle produces and so it cools all the sides of the wax, tallow or fuel, as to keep the edge much cooler than the part within; the part within melts by the flame that runs down the wick as far as it can go before it is extinguished, but the part on the outside does not melt.

"But how does the flame get hold of the fuel?" he puzzled. "There is a beautiful point about that—capillary attraction." Faraday entered into a dialogue with his audience—in this case, mainly children of about 11 years old. "Capillary attraction!" you say—"the attraction of hairs . . . . It is that kind of action which makes two things that do not dissolve in each other still hold together," he explained. And to keep his young audience involved in the conspiracy of learning, there was advice on how to flood the bathroom and establish an alibi at the same time: fill the basin with water, drape a towel over the edge, and capillary action would see to the rest while you innocently got on with your homework downstairs. "The Chemical History of a Candle" was full of such tricks and anecdotes, of "clever costermongers" and prunes in brandy. Nonetheless, this was serious business; Faraday had promised to show how the candle was an "open door" to science.

Science was firstly beautiful, powerfully beautiful. "A combustible thing like that, never being intruded upon by the flame is a very beautiful sight; especially when you come to learn what a vigorous thing flame is—what power it has of destroying the wax itself when it gets hold of it," Faraday enthused. Not for him the lament that scientific understanding lessened the appreciation of natural beauty—understanding enhanced his enjoyment. And, in near-biblical terms, this deeply religious scientist continued: "You have the glittering beauty of gold and silver, and the still higher lustre of jewels . . . but none of these rivals the brilliancy and beauty of flame . . . . The flame shines in the darkness, but the light which the diamond has is as nothing until the flame shine upon it."

The candle also enabled Faraday to take his audience into the scientific method and the general laws it uncovered. We have already seen how he explained the formation of the candle's "cup." Going into this more deeply, Faraday showed:

> If I made a current in one direction, my cup would be lop-sided and the fluid would consequently run over—for the same force of gravity which holds worlds together holds this fluid in a horizontal position . . . . We come here to be philosophers; and I hope you will always remember that whenever a result happens, especially if it be new, you should say "What is the cause? Why does it occur?" and you will in the course of time find out the reason.

Thus, Faraday took his audience through simple cause and effect to universal gravitation, the general method of inquiry of science, and the optimism that imbued its Victorian practitioners.

In Faraday's time, chemistry was a subject that held endless promise. The dreams of the alchemists were being realized: new materials, colors, smells, and tastes were at the disposal of the chemist, and familiar substances were being analyzed to reveal their unfamiliar components. Maybe it was because he lived in an age of "chemical innocence"—a paradise in which pure "nature" had yet to give way to a tainted and politically charged "environment"—or maybe he was just good at it, but Faraday had a knack of dealing with his subjects in as straightforward and everyday a way as possible. Look at how Faraday dealt with the candle flame itself:

> It is a bright oblong . . . and besides the wick in the middle, certain darker parts towards the bottom, where the ignition is not so perfect as the part above . . . . How remarkable it is that that thing which is light enough to produce shadows of other objects can be made to throw its

> own shadow . . . . Curiously enough . . . what we see in shadow as the
> darkest part of the flame is, in reality, the brightest part.

In a later lecture, Faraday investigated the flame in greater depth, showing
that vapor drawn off from the dark inner part of the flame could itself be made to
ignite. No such vapor came from the brightest part of the flame. Instead, soot particles could be detected. "You would hardly think that all those substances which
fly about London, in the form of soots and blacks, are the very beauty and life of
the flame," he said.

The chemist and author Peter Atkins has pointed out that a modern version
of Faraday's lectures could use the candle to probe the latest scientific understanding: "The region of the incandescent flame is the region which epitomizes
the content of quantum theory, a twentieth-century paradigm of science; perhaps
*the* paradigm of science for this century." But the gulf between lecturer and audience is much wider now, as Atkins explains. "Here is a major region of modern
research, which has become open to investigation only with the availability of
modern computers."[4] Midway, in both time and unfamiliarity, between Atkins's
full quantum-mechanical treatment of the candle and Faraday's, chemist Geoffrey
Martin introduced his audience to molecules and the kinetic theories that govern
their behavior:

> Thus the heat and light of our gas or candle flames are due to the clash-
> ing together of oxygen of the air and the constituent molecules of our
> gas and candles . . . . The chemical reactions which occur within it are
> of such complex nature and succeed each other so rapidly that chemists
> are even now uncertain as to exactly what is going on inside it.[5]

Comparing the language used by these three authors is enough to get a feel
for the increasing complexity with which chemists have viewed their world over
the last one-and-a-half centuries. This complexity comes certainly in part from the
increasing depth to which chemists have investigated the materials and reactions
that interest them. But doing it has distanced them more and more from daily experience: what has been hardest won for the chemists—the knowledge they are
most proud of—also makes them more and more remote from their public audience. And, faced with this, it seems that many chemists have turned their back on
the public and on Faraday's example and have retreated into the obscurantism of
symbolic formulas and runic diagrams.

We may also ask how much this retreat has been in response to changing
public attitudes. As we discussed in Chapter 2, World War I appears to have been
something of a watershed in the popularity of chemistry. It is often known as "The

Chemists' War"; but in the aftermath of the carnage, when images of blinded, gasping young victims of poison gas attacks staggering aimlessly in the mud or being led away from the front in pathetic files had burned their way into the public consciousness, association with the war was the last thing chemists wanted. At the 1924 British Empire Exhibition, leading British chemist E. Frankland Armstrong was forced to admit that there was public sentiment against science, quoting John Galsworthy to the effect that chemists were "more hopeful of perfecting poison than of abating coal smoke." Armstrong argued:

> This view is surely based on a misconception, due to the fact that we view the happenings of the world today in a mirror prepared by a Press which only reflects the sensational. Scientific work leading to social progress is going on apace; still it is only when it approaches the sensational that it becomes good copy.[6]

These are sentiments with which many of today's chemists sympathize: in the public mind, even the word "chemical" often has connotations of unnatural tampering, pollution, and risk. Chemist and author John Emsley points out in his *Consumer's Good Chemical Guide* that people happily use "chemicals" every day, for their food, cosmetics, medicines, and clothes.[7] Thus, Emsley has returned to the technique of popularizing chemistry by showing its day-to-day relevance. But his is also a political message: his book "is about chemicals that are commonly regarded as threatening. I hope I will be able to persuade you that some of them are not—that they are good rather than bad." Chemistry, as a whole, lives in such hope.

## Simply Beautiful—Why $C_{60}$ Is Such a Lovely Molecule

For the mass media, "good stories" in chemistry may be rare, but they are not unheard of. On October 10, 1996, newspapers around the world announced that Harry Kroto, of the University of Sussex, and Bob Curl and Richard Smalley, of Rice University in Houston, had won the 1996 Nobel Prize for chemistry for their elucidation of the structure of a new form of carbon. This third form is known variously as $C_{60}$, buckminsterfullerene, or soccerene, on account, of, respectively, its empirical chemical composition, the resemblance of its molecular structure to geodesic domes designed by the architect Buckminster Fuller, and the fact that its "molecular cage" models, which look like soccer balls, can be kicked around the research lab without coming to harm. As well as newspaper and magazine articles, two popular books were written about $C_{60}$: *Perfect Symmetry* by Jim Baggott and *The Most Beautiful Molecule* by Hugh Aldersey-Williams.[8] It was the subject of a BBC television program, "Molecules with Sunglasses" in the "Horizon" series. Kroto's own film company, which he set up with a view to popularizing science, also produced a video of him lecturing on $C_{60}$ at the Royal Institution.[9]

So it seems that $C_{60}$ made a "good story" about science. But just what makes a good science story for the public is an open question, although some consensus does emerge from this particular debate.[10] For a start, there has to be something new, preferably something unexpected. The $C_{60}$ story certainly was that: a new form of carbon that not even the scientists who designed the experiment had expected. When Kroto first got together with Curl and Smalley, they planned to make long chains of carbon atoms all joined together, such as might be found in interstellar gas clouds. But their experiment at Rice University in August 1985 indicated unusually stable and abundant "clusters" of carbon that were a surprise to everyone. Only carbon had gone into Smalley's equipment, and so only carbon could come out, but neither of the two known forms of carbon—black graphite, soft and familiar in the form of pencil lead, and diamond, translucent and extremely hard—fitted the bill. If the experiment proved to be repeatable and the scientists were interpreting the results competently, the textbooks of the future would have to tell students that there were three forms of carbon, not two.

Scientific research makes a good story if it can be followed both as science and as narrative at the same time.[11] The $C_{60}$ story had the plot-line of an "Indiana Jones" adventure story: the heroes began their adventure by looking for one thing—long chains of carbon that might be of academic interest to astronomers—and found something infinitely more precious: a completely new and unsuspected form of carbon. The treasure was in their hands. But, just as for Indiana Jones, the adventure would be about bringing it safely out of the jungle of controversy and home to the museum of acceptance for all to admire. Just three months after starting their work, the $C_{60}$ team announced their findings to the scientific community in terms that made it clear they did not expect an easy ride. In November 1985, the work was published in *Nature*. The paper, "$C_{60}$: Buckminsterfullerene," by H. W. Kroto, J. R. Heath, S. C. O'Brien, R. F. Curl, and R. E. Smalley, is just two pages long, and, as Aldersey-Williams demonstrates, it is a fine example of scientific rhetoric.[12] Even its title "says all that needs to be said and yet says it in such an intriguing way that the casual reader—and, importantly in a general science journal such as *Nature*, the non-chemist—is drawn in, wishing to learn more."

After explaining how they justified their proposal for the existence and structure of $C_{60}$, Kroto et al. supplied the motivation for fellow scientists to pay them due attention:

> Assuming that our somewhat speculative structure is correct, there are a number of important ramifications arising from the existence of such a species. . . . Because of its stability when formed under the most violent conditions, it may be widely distributed in the Universe.

From a "somewhat speculative structure" to cosmic ubiquity—if you are going to propose something that breaks all the rules of organic chemistry, you might as well do so on a grand scale. "Threshold" is as important a concept in science as it is in the media.

The publication of the *Nature* paper by no means settled the affair as far as the wider scientific community was concerned. In the absence of a sample of *pure* $C_{60}$—although it could be made to predominate in Smalley's apparatus, it was always accompanied by other carbon clusters—there was no chance of obtaining further analyses that would really confirm the closed cage structure that Kroto and his team were proposing. The *Nature* paper attracted some devoted followers to the $C_{60}$ camp, but for five years most chemists thought the new molecule a pretty but fantastical beast; and while they enjoyed the conference presentations made by Kroto's group for their imagination and graphics, many thought such speculative chemistry unwise. The $C_{60}$ scientists struggled to win both the approval of their peers and, in Kroto's case, money to continue the research. (Problems continued right up until the moment that the Nobel Prize was announced. "Ironically," reported the *Guardian* newspaper, "it came only hours after [Kroto] was turned down for government funding for new research into the same subject."[13] An embarrassed research council swiftly reversed its decision.)

The media will generally try to personalize a story, if it can be done. Kroto, Smalley, and Curl were all strong characters, and this comes across in all of the books and programs—when interviewed, each scientist had a different perspective on just who it was who had worked out that 60 carbon atoms could join together to form a closed structure shaped like a soccer ball, and on how the team did it. At the start of the project, the researchers worked closely together. According to Baggott:

> Together they lived and breathed the experiments, sometimes working late into the night . . . . At the end of their long session in the lab, which would sometimes not conclude until two or three in the morning, Kroto, Heath and O'Brien [two of Smalley's research team] would make their way to the House of Pies . . . . Over endless cups of coffee, they talked about art, books, music, science and religion.

But every team, however intrepid, suffers its own internal strains:

> Growing tensions between Kroto and Smalley finally spilled over into confrontation. Despite their outward desire to maintain that the discovery of Buckminsterfullerene had been a team effort, inwardly they had come to disagree bitterly over specific details . . . . Kroto returned to Sussex [in April 1987] despondent but determined to defend his interests against what he saw to be an unjust attack . . . .

Besides the infighting in the $C_{60}$ camp, there were rivals on the horizon. Scientists at the oil company Exxon had discovered carbon clusters like $C_{60}$ in some of their analyses but had paid them little attention. A U.S.–German team, Don Huffman and Wolfgang Krätschmer, had measured a spectrum of $C_{60}$ three years before Kroto and his colleagues but had not realized what they had seen. As it turned out, they got the first pure sample of $C_{60}$, and published its persuasively simple infrared spectrum in *Nature* in 1990. Kroto refereed their paper with mixed feelings: he had been right after all, but he had lost the race to prove it.

The media also like a story to have "relevance" for their audience, even if the scientists are not concerned about the applications of what they have discovered. There were surely more prizes to be had than the one from the Nobel committee: now that $C_{60}$ could be made easily and cheaply, thoughts turned to capitalizing on the discovery. Speculation was rife: for example, why not extend the molecules into sturdy cylindrical cages, fill the cylinders with metal atoms, and revolutionize electronics once again? But when the Nobel Prize was awarded, $C_{60}$ was still waiting to go to work. Kroto did not seem to mind; after all, he had been on his adventure and had already come home with his reward:

> Why should there be a use? We are talking about major shifts in our understanding of nature. We have added a third form of carbon to diamond and graphite. There are huge areas of organic chemistry now opened up that are quite different from anything before.[14]

## BIG SCIENCE, BIG QUESTIONS

### A Relative Revolution

$E=mc^2$ is just about the only equation that manages not to frighten off most of the readership of a popular-science book; and if there is one face that almost everyone recognizes as a scientist, it is the shock-headed, moustachioed visage of Albert Einstein.[15] Einstein first came to the notice of the scientific community in

1905, when he published three papers that would later be viewed as pathbreaking. He accomplished all this, moreover, while working at the patent office in Berne, Switzerland. One of the papers dealt with Brownian motion, the apparently random movement of, for example, specks of dust under the constant bombardment of air molecules. Another paper explained the newly discovered photoelectric effect by treating light as a stream of particles, making use of the quantum relations proposed by Max Planck. The third paper looked at the electrodynamics of moving bodies and outlined the special theory of relativity. Ten years later, Einstein had developed this third strand into the general theory of relativity.

Einstein's new theory proposed that light should be "bent" as it passed close to a massive body, such as a star, by an amount roughly twice that predicted by Newton's theory of gravitation. In 1917, this prediction came to the notice of the British Astronomer Royal, Sir Frank Dyson, who realized that an eclipse due on May 29, 1919, would furnish an ideal opportunity for testing which law of gravitation—Einstein's or Newton's—was correct. Despite the difficulties of the war—not least of which was that Einstein was, at the time, a professor at the University of Berlin in hostile Germany—Dyson organized expeditions to go to the island of Príncipe, off the coast of East Africa, and Sobral, in Brazil, under the leadership of Cambridge astronomer Arthur Eddington. These locations were chosen because the eclipse there would be total. As luck would have it, a group of stars, called the Hyades, would appear close to the Sun in the sky and would be clearly visible during the eclipse. Measuring how much the light from the Hyades was bent—or "deflected," to use the astronomical term—would tell the scientists whether Einstein's or Newton's theory of gravitation was the correct one. The measurements were extremely difficult to do, and there was much scope for error and uncertainty. Only after months of processing and analysis were the results of the expeditions finally reported in London to a joint meeting of the Royal Society and the Royal Astronomical Society on Thursday, November 6, 1919. Einstein, it appeared, was right.

The next day, *The Times* of London announced a "Revolution in Science: New Theory of the Universe: Newtonian Ideas Overthrown." Two days later, the *New York Times* hailed the results as "epochmaking." A media bandwagon, with Einstein the sometimes reluctant, sometimes all-too-willing, passenger, had started to roll, and Einstein was pursued by journalists in a manner usually reserved for film stars and aberrant politicians. How this bandwagon ran, and the effect it had on Einstein and his family over many years, has been well documented.[16] However, sociologists Harry Collins and Trevor Pinch have pointed out that the results from the eclipse expeditions were not as conclusively in favor of Einstein's theory and against Newton's as is often stated in popular accounts and textbooks and that there was room for considerable debate among scientists about the meaning and value of the observations.[17]

In this section, we look at the role played by *The Times* in the first few weeks after the November 6 meeting as it championed Einstein and divided scientists into insiders and outsiders in the debate. We will also examine the claim by Collins and Pinch that those outside of a select band of consenting astronomers were never given the full story of the eclipse expedition and its results, and we will discuss reasons why Einstein and relativity became so popular so quickly.

The article in *The Times* of Friday, November 7, 1919, opened soberly enough: "Yesterday afternoon . . . at a joint session of the Royal and Astronomical Societies, the results obtained by British observers of the total solar eclipse of May 29 were discussed." But controversy was soon in the air: "The greatest possible interest had been aroused in scientific circles by the hope that rival theories of a fundamental physical problem would be put to the test . . . ." So they were, and, in the opinion of *The Times*, the matter was settled: "It was generally accepted that the observations were decisive in verifying the predictions of the famous physicist, Einstein." The article quoted the president of the Royal Society, J. J. Thomson, who, after Dyson and Eddington had spoken, said that the assembled scientists had just listened to "one of the most momentous, if not the most momentous, pronouncements of human thought."

Readers of *The Times* were told that the matter hinged on the amount that the position of bright stars, visible close to the position of the Sun during the eclipse, appeared to have shifted, a process known as "deflection." Reporting Dyson's contribution to the meeting, the article said: "He convinced the meeting that the results were definite and conclusive. Deflection did take place, and the measurements showed that the extent of the deflection was in close accord with the theoretical degree predicted by Einstein, as opposed to half that degree, the amount that would follow from the principles of Newton." As a result, *The Times* concluded, "the qualities of space, hitherto believed absolute, are relative to their circumstances. Space was "warped," *The Times'* reporter explained.

The paper's editors were also deeply moved. In an editorial grandly entitled "The Fabric of the Universe," they thundered:

> From Euclid to Kepler, from Kepler to Sir Isaac Newton, we have been led to believe in the fixity of certain fundamental laws of the universe . . . . But it is confidently believed by the greatest experts that enough has been done to overthrow the certainty of ages and to require a new philosophy of the universe, a philosophy that will sweep away nearly all that has hitherto been accepted as the axiomatic basis of physical thought.

It was to be no passing fancy. Coverage similar in tone and extent continued for three weeks. On November 8, the paper canvased the "views of eminent physicists," since "wide interest in popular as well as in scientific circles had been created by the discussion . . . on the results of the British expedition . . . to observe the eclipse of the Sun on May 29." Parliament was abuzz. "Sir Joseph Larmor FRS, MP for Cambridge University, said he had been besieged by inquiries as to whether Newton had been cast down and Cambridge 'done in'." Charles Davidson, a member of the eclipse expedition, was given space to deal with anyone doubting Einstein, though he warned: "Professor Einstein's theory . . . demanded a good deal more of the dimensions existing in space than can at present be mathematically proved."

*The Times* also gave its readers a pen-portrait of the perhaps not yet so "famous" physicist at the center of attention. "Dr Albert Einstein . . . is a Swiss Jew, 45 years of age"; "a man of liberal tendencies," he was "one of the signatories of the protest against the German manifesto of the men of science who declared themselves in favor of Germany's part in the war." (Less than a year after the signing of the armistice, it was important to show that Einstein was not really German at all.) Einstein's support for Zionism and for the proposed Hebrew University at Jerusalem were favorably commented on.

As the days went by, everyone, it seemed, had something to say about relativity. Cardinal Mercier, interviewed for *The Times* of November 12, "indicated that he would like upon another occasion to discuss Professor Einstein's theories of the universe to which *The Times* drew special attention in a leading article last Friday. 'I cannot go into such deep waters now,' he said, 'but my belief is that Professor Einstein's theories were anticipated some 12 or 15 years ago by Professor Paul Mansion in the Revue Neo-Scolastique which we published at Louvain.'" In an editorial a few days later, the paper mused on the philosophical implications of the new physics: "Observational science has in fact led back to the purest subjective idealism, if without Berkeley's major premise, itself an abstraction of Aristotelian notions of infinity, to take it out of chaos." Continuing with philosophy, a long letter from H. Wilden Carr of King's College London pointed out that the idea of absolute space and time was itself a product of the Newtonian revolution. "The theory of relativity is a return to [Descartes] . . . . What is surprising is that so few, even among philosophers, are mindful how modern the notion of absolute space and time is." On December 2, *The Times* ran a humorous column under the heading "The Run Home, an Einsteinian at Large," in which a tired country clergyman was brought up to date by an enthusiastic monk on a "motor-omnibus." "We live . . . in a curiously curved Space–Time continuum," the chirpy cowl informed the bemused dog-collar.

*The Times'* pièce de résistance came on November 28, when it ran a signed article by Albert Einstein himself. The paper was beside itself with smugness:

> We publish today . . . an article written for our readers by Dr Albert Einstein, whose momentous theory of relativity received the strongest confirmation from the results of the last English Solar Eclipse Expedition. We cannot profess to follow the details and implications of the theory with complete certainty [but] with all respect to the ripe wisdom of our esteemed correspondent, Professor Case, we cannot accept his view that there are still left to us the conceptions of absolute space . . . and absolute time . . . .

With a touch of the humor that was to endear him to audiences across Europe and America in the months to come, Einstein opened his article by remarking: "The description of me and my circumstances in *The Times* shows . . . an imagination on the part of the writer. By an application of the theory of relativity to the taste of readers, today in Germany I am called a German man of science and in England I am represented as a Swiss Jew. If I come to be regarded as a bête noire, the descriptions will be reversed . . . ." (The paper allowed Einstein his little joke at their expense, pointing out, in return, that he had failed to supply an *absolute* description of himself.) Einstein also took the opportunity to make clear his feelings about the recent war. "After the lamentable breach in the former international relations existing among men of science, it is with joy and gratefulness that I accept this opportunity of communication with English astronomers and physicists."

The article, which demanded a high degree of scientific literacy from its readers, then set out the principles of the new theory.

> Since the time of the ancient Greeks it has been well known that in describing the motion of a body we must refer to another body . . . using a system of co-ordinates. Every law of nature which holds with respect to a co-ordinate system K must also hold good for any other system K' provided that K and K' are in uniform movement of translation. [This was Einstein's first principle of relativity theory.] The second principle on which the special theory of relativity rests is that of the constancy of the velocity of light.

Einstein then explained how general relativity worked when considering the correspondence of inertial and gravitational forces. "In the generalized theory of relativity, the doctrine of space and time, kinematics, is no longer one of the absolute foundations of general physics." Around massive objects there was a "warp in space." Unlike *The Times* itself, Einstein was careful to play down the revolutionary nature of his theory. "No one must think that Newton's great creation can be overthrown in any real sense by this or any other theory. His clear and wide ideas will forever retain their significance as the foundations on which our modern conceptions of physics have been built."

In the three weeks since first reporting the November 6 meeting, *The Times* had been far from acting as a neutral commentator on the discussions taking place within the scientific community. Those who were unconvinced by the results of the eclipse expedition got short shrift from the paper. Prior to the expeditions to Príncipe and Sobral, physicist Sir Oliver Lodge had poured scorn on such undertakings. In its November 7 report, *The Times* was quick to remind him of this: "It is interesting to recall that Sir Oliver Lodge, speaking at the Royal Institution last February, had also ventured on a prediction. He doubted if deflection would be observed, but was confident that if it did take place, it would follow the law of Newton and not that of Einstein." So would Sir Oliver take up the challenge of Dyson and Eddington? *The Times* reported that, after Dyson and Eddington had spoken, J. J. Thomson said that he "was confident that the Einstein theory must now be reckoned with, and that our conceptions of the fabric of the universe must be fundamentally altered." "At this stage Sir Oliver Lodge, whose contribution to the discussion had been eagerly expected, left the meeting," the paper sneered.

All that was left to poor Sir Oliver was a lame explanation on the letters page to the effect he had left in order to catch a train to Bristol. Other scientists with ideas opposing Einstein, or even just offering alternative interpretations, were similarly confined to writing letters, while Eddington, playing Einstein's "bulldog" just as T. H. Huxley had done for Charles Darwin, was invited to write signed and prominent articles. Just a week after the November 6 meeting, *The Times'* editorial had decided that "Every authority is agreed on the interpretation of the results . . . . The deflection . . . corresponds with the prediction made by Dr Einstein from his theory of relativity." If you were not "agreed," you were not an "authority"—*quod erat demonstrandum*.

But did *The Times* play fair by its readers? Collins and Pinch claim is that the public never got to know about the magnitude of the uncertainty surrounding Eddington's eclipse measurements. Their argument hinges on the fact that only some of the results supported Einstein's theory and that Eddington then produced *a posteriori* reasons for rejecting those that did not give the required deflections. In all, three telescopes were taken to make measurements, one to Príncipe and two to Sobral. The data obtained in Príncipe were, in Eddington's own words, "meagre" because of poor weather. But what was obtained did show that the apparent positions of the Hyades stars had been deflected so that their "new" positions lay more or less on a straight-line graph with a slope predicted by Einstein's general relativity.

At Sobral, where observing conditions were much better, there was another telescope of the same type as that taken to Príncipe. Results from this telescope, however, showed that the deflections of the Hyades were much more in line with Newton's theory. According to some scientific criteria, the greater amount of data should have carried more weight in deciding which theory had been verified. But Eddington argued that the heat at Sobral had "warped" the plates, and he discounted the data obtained from this telescope. Instead, he made great play of the data from the second telescope that had been taken to Sobral, which, though limited in quantity, were felt to be of a much superior quality. These data supported Einstein. Nonetheless, there was room for doubt in the data taken as a whole, and this was expressed and discussed in a paper written by Dyson, Eddington, and Davidson for the *Philosophical Transactions of the Royal Society*.[18] Collins and Pinch argue that the public never got to hear of this uncertainty.

Certainly, only scientists would have read the paper in the *Philosophical Transactions*. Readers of *The Times* were not told that some of the data appeared to support Newton but Eddington and his co-workers had rejected these after the event. In that sense, Collins and Pinch are right. But there was another source of information on the eclipse expedition. On his way back from Príncipe, Eddington had started writing a book about Einstein's theories. *Space, Time and Gravitation* was published in 1920, and Eddington's aim was to popularize relativity theory to the scientific community beyond the narrow circles of astronomy and electrodynamics.[19] The book was also read by people outside of the circles of science, many of whom would have struggled through Einstein's own article in the *The Times* and been interested enough to read further. *Space, Time and Gravitation* included a highly accessible chapter entitled "Weighing Light," in which Eddington went into considerable detail about the eclipse expedition and his reasons for accepting the data that confirmed Einstein's theory and for rejecting those which appeared to favor Newton's. Figures that appeared in the *Philosophical Transactions* paper were also published in "Weighing Light." Moreover, almost all of the information concerning the uncertainty in the data taken at the two observing sites that appeared in the *Philosophical Transactions*, used by Collins and Pinch in their critique, can also be found in "Weighing Light." So members of the public interested enough in the new physics to read this book were given more or less the same information as the scientists reading the *Philosophical Transactions*; the "interested public," at least, if not the public generally, did get the inside story.

It is also interesting to ask why was there such a large public for relativity. Physicist Abraham Pais has argued that language was an important factor:

> Even though so very few had and have a real grasp of Einstein's contributions, their contents nevertheless lend themselves very well to expression in everyday language. Take "Space Warped." Everybody thinks he or she knows what "space" means, and everybody knows what a warp is. Yet hardly anyone understands the meaning of warped space. At the same time, the words are sufficiently familiar to convey to the reader a sense of having been let in on a highly esoteric piece of information.[20]

The appeal of the "familiar mystery" has never waned. At the time, however, the apparent familiarity of the language also gave rise to much confusion. "Widely believed misconceptions arose: that Einstein taught everything is relative, including truth; that all observations are subjective; that anything is possible."[21] So factors other than language must also have been at work.

Historian of modern physics Arthur I. Miller has recently pointed out that the early 20th century was a time in which fundamental ideas about space and time came under critical examination by the likes of Henri Poincaré in mathematics, Henri Bergson in philosophy, Marcel Proust in literature, and Georges Braque and Pablo Picasso in art. There was a *Zeitgeist* in which these and many others questioned the very fabric of the world from every possible angle.[22] So what was the *Zeitgeist* of the immediate post-World War I years? *The Times'* report of the November 6 meeting came out just two years to the day after the Bolshevik Revolution in Petrograd. *The Times'* headline promised a "revolution in science." Einstein was depicted as "sweeping away" the old order of physics, an image redolent of contemporary propaganda posters of Lenin sweeping the old social order off the globe. The paper's other columns carried reports of negotiations between government ministers and union leaders representing the miners, part of a long-running dispute that was to culminate in the British General Strike of 1926. The pen-portrait of Einstein remarked favorably on his support for the Spartacist uprising in Germany, which ended with the brutal murders of its leaders, Rosa Luxemburg and Karl Liebknecht.

So, did Einstein's "scientific revolution" mirror those going on in contemporary society? Was part of the popularity of relativity due to this *Zeitgeist*? If so, *The Times*, mouthpiece of the British Establishment, was not the most obvious ally of abrupt and—intellectually, at least—violent change. Maybe the paper played the key role that it did in backing relativity, and thus ensuring that Einstein became a household name, for the simple reason that, by methods not immediately recognizable as scientific, it just got it right.

## Ripples from the Edge of Time—a Case of "Hype" or a "Good Story"

The 1919 press coverage of Einstein and the eclipse expedition shows how the press can intervene in a scientific debate and publicly "settle" issues still uncertain among the scientists themselves. The media worked to their own agenda, perhaps at the expense of due scientific process. Our next example is of the coincidence and convergence of scientific and media debates, and concerns the reporting in 1992 of the results from a rather obscure NASA satellite called CoBE, the Cosmic Background Explorer. On April 23, 1992, the team of scientists working on CoBE told the American Physical Society that they had, at last, detected "ripples" from the early universe which eventually gave rise to the galaxies, stars and planets we see today. Their announcement made front-page headlines around the world and was a major item on television and radio news programs—coverage that was all the more remarkable given that the satellite neither offered hope to humanity of a cure for a serious illness nor came up with results that revolutionized science overnight. CoBE merely told astronomers what they, by and large, already believed: that the universe did begin with a big bang. In order to understand the extent to which the CoBE story dominated the media on the day, and continued to interest them for many months to come, we must look at the tensions that existed in the scientific community and at how the media, in Britain at least, had for some time prior to April 1992 been debating the role of science in answering some of the big questions of our existence and our times. The difference, it seems, between an insignificant ripple and a big splash is timing.[23]

Modern big bang cosmology rests essentially on two pieces of evidence. The first is the observation made by the astronomer Edwin Hubble in 1929 that distant galaxies are moving away from us and are doing so at a speed that increases the further away they are. The conclusions that follows from this are that the universe is expanding uniformly, and that if one runs the clock of cosmological history backward over billions of years, at some time in the distant past everything in the universe must have been so close together that it formed a single point—a singularity in space and time. The second piece of evidence is the microwave background radiation, with a temperature of 2.73 kelvins, coming from every point in the cosmos. Astrophysicists interpret this radiation as the dying embers of the fireball in which the universe, as we know it, came into being. Conceptually, the hot big bang scenario for the origin of the cosmos is very simple. But problems emerge when the laws of physics are employed in a detailed analysis of the current universe. For a start, the cosmic microwave background radiation is very smooth—it has the same "temperature" no matter where one looks. That implies that during a crucial phase in the life of the early universe, all parts of it were able to communicate with one another. The only way that could have happened is if one tiny part of the universe had expanded with enormous rapidity and crowded out the rest. But this process of cosmic "inflation" has, like its economic counterpart, serious consequences. Inflation requires the existence of dark matter, composed, in all probability, of exotic particles that interact with "normal" matter so rarely that no one has ever detected them. Large amounts of dark matter are needed, too: as much as 90 percent of the universe may be composed of this intangible stuff. Another problem was that, prior to 1992, all the measurements of the temperature of the background radiation from different directions indicated that it was too perfectly uniform—smooth to at most a few parts per million. If, in the early universe, all the fluctuations and unevenness had been ironed out by inflation, how did we ever get to today's universe of lumps—lumpy galaxies, stars, planets, and people?

It was to address this latter question, in particular, that CoBE was launched in 1989. Its instruments were designed to "map" the temperature of the background radiation from all parts of the sky in an attempt to pick up tiny fluctuations from the average. The microwave background radiation does not date all the way back to the big bang itself, but to some 300,000 years after the primal fireball, when the temperature of the universe had dropped to around 6000 kelvins and neutral atoms could form out of the sea of protons and electrons. Fluctuations would have to be present in the background radiation if the cosmos, according to cosmologists' models, were to form galaxies and other structures in the 10 to 15 billion years since the big bang. But CoBE's three years of data collection and analysis were nervous ones for the cosmological community. "CoBE's sensitive instruments listened carefully for the faint whisper of the cosmic explosion . . . . For more than two years, the satellite found nothing. There were jittery mutterings among scientists," wrote former *New Scientist* journalist Marcus Chown.[24] He cited CoBE team member Dave Wilkinson: "If CoBE gets to one part in a million and still sees the sky smooth, big bang theories will be in a lot of trouble." Fellow team member George Smoot confirmed this view: "By early 1992, the continued inability to detect wrinkles in the background radiation had become a serious embarrassment for . . . big bang theory."[25] Within the world of astronomy, there were further concerns in the years leading up to 1992. *Nature* ran a number of articles that questioned the existence of dark matter, at least in its most theoretically favored cold form. There were problems too surrounding the value of the Hubble constant, $H_0$, which measures the rate at which the universe is expanding and, inversely, its age. Large values of $H_0$ were being claimed, which made the universe younger than some of the oldest stars to be found within it—clearly an impossibility, but one that cast doubt on one of the two main observational supports for standard cosmology. And a team of rebel astronomers, including veteran big bang antagonist Fred Hoyle, was given space to question the whole basis of big bang cosmology in *Nature*'s prestigious pages, albeit under the unusual heading of "Hypothesis," and not without a reply from leading orthodox cosmologists.[26] The tension was mounting.

The public debate as to whether science has anything to say about the existence or otherwise of God and about the meaning of life has always been with us; but, toward the end of the 1980s it intensified in Britain, particularly after the publication of Stephen Hawking's *A Brief History of Time*, which concluded with the much-quoted suggestion that answering the question of why we and the universe exist "would be the ultimate triumph of human reason—for then we would know the mind of God."[27] The debate was carried forward in the form of books written by scientists, philosophers, and social commentators and was often aired in the national media. Shortly before the release of the CoBE results, the *Daily Telegraph*, in particular, had been using its science page to air opposing views regarding the nature, worth, and meaning of science. Those "against" science included novelist Fay Weldon, who attacked scientists for their failure to answer questions any 6-year-old would want answered (such as, according to Weldon, "what is the ghost in the machine which is the self?"), and columnist Mary Kenny, who lambasted scientists for their lack of "soul." Defending science were embryologist Lewis Wolpert, who was outraged at the growing "anti-science bandwagon," and chemist Peter Atkins, who was proud to be part of the drive to replace superstition and ignorance by a scientific understanding of the world, especially if this might also undermine religious belief.

Other newspapers, too, carried articles. The *Guardian* featured cosmologist Paul Davies's explanations of the religious insights provided by modern science. The *Financial Times* ran a modern cosmology version of Genesis:

> In the beginning God created the laws which govern the heaven and the earth. And the earth was without form and void. And the void was filled with a ferment of particles appearing and disappearing in the darkness. And God created quantum mechanics, relativity, quarks and

superstrings, each after its kind. And God said let there be light and there was a very Big Bang . . . .[28]

In April 1992, the annual Edinburgh Science Festival hosted a debate between geneticist Richard Dawkins, who claimed that "faith was a cop-out," and John Habgood, the Archbishop of York, who felt God was unmoved by whether there had or had not been a big bang. This debate was carried at great length in two issues of the *Independent*, which coincidentally straddled the announcement of CoBE's results. And the day after CoBE made the front pages of all of the British broadsheets, Bryan Appleyard, whose views are discussed in Chapter 3, used the pages of *The Times'* Saturday magazine (which had gone to press before the CoBE story broke) to round on the growing "scientism" of leading scientists. In this already charged media environment, suddenly there was news from CoBE.

"Co-opting" a story—linking it to one that is already running—can add enormously to the story's news value and enhance its media coverage (see the discussion of news values in Chapter 5). Sometimes this co-option is coincidental; sometimes one party or another engineers connections with the aim of making a bigger splash. The news that CoBE *had* found fluctuations in the background radiation temperature of the order of a few parts per million was due to be announced at a meeting of the American Physical Society in Baltimore on April 23. NASA had arranged a press conference immediately afterward, to be headed by their CoBE team leader, John Mather. But, according to Chown's version of events, the Lawrence Berkeley Laboratory in California was concerned that it would get insufficient credit for the results. So it issued its own press release, which hit the agency wires before the official press conference. As a result, by the time that the CoBE team addressed the American Physical Society, the BBC had already run the story on its radio and television news broadcasts, and Lawrence Berkeley's George Smoot had been fixed in the public mind as the key person.

Smoot made sure that he stayed in pole position. In his address to the American Physical Society, he claimed that the English language had too few superlatives to express the importance of what he was to announce. In the press conference afterward, Smoot sought to explain the significance of CoBE's results by saying that it was like "seeing the face of God." Whether he had taken a leaf from Hawking's book or whether he was simply reacting to the excitement of the moment and the collective sigh of cosmological relief, Smoot gave the media the opportunity they needed to "co-opt" CoBE into the "big questions, big physics, big answers" debate. Nor was Smoot alone. British cosmologist Michael Rowan-

Robinson describes the media feeding frenzy that followed: "That Thursday [April 23] was an especially testing day, since astronomers had to respond to journalists off-the-cuff without knowing the details of the CoBE announcement . . . ." Stephen Hawking was quoted as saying that CoBE's results were "the discovery of the century, if not of all time."[29] Mather himself said they had found "the missing link of cosmology." And Chicago cosmologist Michael Turner ensured that he fitted into the frame by praising Smoot and Mather for finding "the Holy Grail of cosmology."

Another news value on which CoBE scored well was "facticity." The press went to town, giving their readers a brief (and, in some cases, not so brief) lesson in cosmology. The facts offered to readers ranged from the age of the universe (15 billion years) and the "size" of the ripples detected by CoBE to details about the satellite and its orbit. Scientists often have a horror of dealing with the more popular newspapers, fearing that, whatever they say, the journalist will either get it wrong or simplify it to the point of banality. True, the most "factual" of the British papers on Friday, April 24, was one of the quality broadsheets, the *Independent*. But the middlebrow *Daily Mail* scored as well on factual content as the prestigious *International Herald Tribune*, which had run the Press Association's copy, by making use of signed articles by Hawking and TV astronomer Patrick Moore. And the tabloid *Sun*, under the tongue-in-cheek headline "We find the secret of creation…and it only took 15 billion years," scored better than the science-loving *Telegraph*. With the exception of the *Daily Express*, which thought the universe as a whole had started only 300,000 years ago, the dailies had covered the science of CoBE in considerable detail, and accurately.

Some, too, either had made use of the "photograph" of CoBE's ripples supplied by NASA or had commissioned their own graphics to assist in explaining the basic concepts of cosmology. The story had broken too late for the specialist science weeklies—*Nature*, *Science*, and *New Scientist*—to run anything that week, but in the following issues they had little to add to the factual content of the discussion that had not already appeared in the popular press.

Once the basic facts had been dealt with, the door was open to explore the significance of the new results, in the same way that *The Times* had dealt with Einstein's theory. Throughout the weekend, and, intermittently, over the next few months, articles that made reference to CoBE appeared as part of the ongoing discussion of the importance of modern science. Typical of articles discussing science and religion was one in the *Sunday Times* by the Reverend John Polkinghorne, a former physicist. Polkinghorne claimed that science and religion dealt with different aspects of knowledge:

> Science asks how things happen, theology asks why. Faith is not a matter of choosing between science and religion, of believing in God the Creator or the Big Bang. The latest [CoBE] discovery is very interesting and exciting, but it has no implications for God.[30]

Other writers disagreed. Tom Wilkie of the *Independent* argued:

> Not since T. H. Huxley debated evolution with "Soapy Sam" Wilberforce, in 1860, has there been such intellectual and popular interest in the claims of science . . . . But while cosmologists now give confident accounts of the creation and history of the universe, doubts have been creeping in as to whether, if Science has dethroned God, It is big enough to take His place.[31]

On a lighter note, the *Daily Mail* followed its detailed coverage by asking whether CoBE had given God his P45 (the British pink slip). In a similar vein, *The Times'* columnist Bernard Levin found "reports that creation had been explained much exaggerated" and chuckled:

> Well, poor old God. Hardly had he recovered from the news that he was a hallucinogenic mushroom, than his most cherished secret—how the universe came into being—has been revealed . . . my best guess is that God has a sense of humour, and at this moment is rolling on the floor with a handkerchief stuffed into his mouth and tears pouring down his cheeks.[32]

The extended run given to the CoBE story, as a result of its co-option into the science/religion debate, gave journalists the opportunity for personalization. Stephen Hawking was widely profiled; Hawking's intellectual quest to "know the mind of God" was contrasted with his physical appearance "crumpled like a broken doll in a wheelchair, imprisoned by motor neurone disease within the wreckage of a human body."[33] Previously unknown to the public, George Smoot almost proved too much for the *Independent*'s Sara Wheeler. "The annoying thing is that besides being one of the greatest astrophysicists he is also modest, charming, funny and nice. Worse still, he is handsome," she gasped.[34]

By and large, the extensive media coverage given to CoBE, its team members, and associated cosmologists was very positive, when it was not being deliberately satirical. But one sour note was sounded early on. Writing in the *Observer* the Sunday after the NASA press conference, Robin McKie raised publicly what some journalists were wondering in private. "Many newspaper readers, especially those rendered dizzy by the convoluted explanations which the probe spawned last week, could reasonably ask: 'Was it all hype, or did CoBE really return breathtaking data?' Predictably, the answer is: both."[35]

McKie's question was asked again in the BBC television documentary "Breakthrough or Ballyhoo?", hosted by Richard Smith, editor of the *British Medical Journal*. The program highlighted one matter that had not been made clear to the public: the "ripples" shown in the pictures released by NASA were not ripples in the background radiation, but were mainly noise. It took a complicated statistical analysis presented in three papers in the specialist *Astrophysical Journal* to show that, beneath this noise, CoBE's data contained genuine cosmological ripples.[36] In releasing the CoBE pictures and allowing the media to believe that what they were looking at was the structure of the early universe, NASA had been somewhat economical with the truth. Nor had CoBE delivered *the* answer, as the media had claimed. Indeed, there were complaints that, far from narrowing down the possible models of the universe, CoBE had actually opened the door to several more variations on the big bang/inflation theme.[37]

Nevertheless, when the opportunity arose to revisit CoBE in reviews of books by Smoot, Chown, and Rowan-Robinson in the autumn of 1993, the scientific establishment generally agreed with the importance its results had been accorded; the media, too, were happy with their performance.[38] But there was an indication in the press that claims of cosmological proportions would not always get a smooth ride. On November 18, 1993, the *Guardian* prominently reported that a British astronomer was claiming to have solved the riddle of dark matter, "the biggest mystery in the universe." Three days later, the *Sunday Telegraph* slammed the *Guardian* for falling for "space hype." Cosmology is big, and the media are prepared to run it big—but not every time of asking.

## OILING THE ENVIRONMENT

The environment is a very complicated place: most branches of science have something to say about it, and environmental science itself is still taking its first tentative steps toward the complex syntheses of knowledges that seem to be required for understanding a relentlessly dynamic nature. Yet despite its complexities, the science of the environment is one that has attracted considerable and often powerful public interest; and since the shock publications and popular protests of the 1960s, and with political and business interests also at stake, the environment has been a staple of science journalism. In what follows we look at stories of oil pollution and ask what scientific literacy might mean for people whose closest acquaintance with environmental science comes through the mass media. We use Durant's framework for scientific literacy[39] (as discussed in Chapter 4) to explore how the public might understand the science of oil and the environment.

On the evening of Thursday, February 15, 1996, an oil tanker ran aground off the coast of Wales. The Liberian-registered *Sea Empress* had been heading for the Texaco oil refinery at Milford Haven: its crude cargo began to leak into the sea, threatening the coastal wildlife nature reserves. Ocean-going tugs raced to the scene, and a clean-up operation, involving planes spraying detergent on the oily waters, began as soon as daylight broke. Divers surveyed the underwater damage to the tanker's holds, attempting to estimate just how quickly the oil was escaping. Reporters, taking their cue from environmental groups, warned that the greatest environmental catastrophe ever to hit Britain's coasts was looming. Ministers were challenged on whether or not they had learned the lessons of pervious maritime oil spillages, and debates raged about tanker design and the effectiveness of the salvage mission. Other media voices countered, however, that previous gloomy predictions of disaster had grossly exaggerated the problem and underestimated the environment's powers of self-restoration.

Reporters and camera crews arrived in West Wales, where they interviewed fishermen, the salvage operators, and representatives of the nature reserves and of environmental groups. Although the story broke too late for there to be much coverage in the next morning's editions, by the weekend the papers carried front-page articles with pictures of the spreading slick. It was the lead item on television and radio news bulletins. For a week, until the *Sea Empress* was finally dragged into Milford Haven, the drama of the battered tanker, struggling tugs, and the treacherous black cargo was prominent in the news pages and broadcasts. By the Thursday after the original accident—February 22—some 70,000 tonnes of crude oil had leaked into the sea, approximately half of the tanker's original load. Above a half-page picture of the crippled *Sea Empress* surrounded by a bevy of heaving tugs, the *Guardian*'s front-page headline ran "The sea that turned to treacle." The paper warned: "Tanker pulled off rocks but marine life is threatened . . . ." The lead article pointed out that in just the last day before its final rescue, the tanker had put 15,000 tonnes of oil into the sea and that the local fishing industry had imposed a voluntary ban on fishing within 40 miles of the affected area: "Phil Coates, a spokesperson for the industry, said that his colleagues were worried that the oil would kill shellfish and affect other catches, and that public confidence in the industry would also be damaged."

Eyewitness Edward Pilkington on Skomer (one of the nature reserves threatened by the oil) spoke of the island as a "haven of tranquillity in an angry sea." Pilkington explained that this would soon change: during the imminent breeding season, Skomer would be "heaving with 14,000 guillemots, razorbills, shags, gannets and puffins." And these creatures were all in danger. "At a distance of half a mile from the *Sea Empress*, the wake of our boat turned suddenly into a bubbling cauldron of treacle," he added. The local Wildlife Trust warned: "The enormous volume of oil . . . will undoubtedly have long term consequences for the islands' fragile ecosystems." Inside the *Guardian*, the debate ranged around whether the British government had or had not acted on the recommendations of the official inquiry into the wrecking of the Braer oil tanker off the Scottish Shetland Islands in 1993.

The issues raised in the *Guardian* also featured strongly on TV news bulletins. The Channel 4 evening news on Thursday, February 22, introduced the item with: "The *Sea Empress* may now be safely tied up but the pollution threat is far from over with oil still leaking." In an on-the-spot interview, local conservationist Jonathan Hughes, pointing across the water to Skomer, said that the sea surface was "revolting" and that oiled seabirds were already being washed ashore. "But the main problem will be under the water, with the crustaceans, the invertebrates and the fans of corals. And that damage will take some time to make itself evident," he warned. Bird rescuer Jean Hains raised the question of tanker design and asked why tankers that did not have two skins were allowed into British coastal waters. On behalf of the—by then—much criticized salvage operation, a spokesman insisted that the *Sea Empress* was now safely docked, and its tanks were being made inert. "The situation is so much under control that there is hardly any chance that we will get further leakage," he said—a claim quickly contradicted by reporter Alex Thompson. Cutting back to the House of Commons, the program showed Transport Secretary Sir George Young assuring Members of Parliament that there would be an inquiry that would look at the salvage operation as well as at the cause of the accident, while the opposition widened its attack to blame the disaster on the depletion of the British merchant shipping fleet. And the government was hit by a final sideswipe from Thompson: "A great many [local] people . . . want decisive action . . . instead they are just getting a lengthy inquiry."

This coverage gave the distinct impression that the government and the salvage company—not to mention Texaco and the owners of the *Sea Empress*—were collectively responsible for Britain's worst ecological nightmare. But that same day, a week after the accident, voices of moderation and downright skepticism were to be found, in the shape of newspaper science journalists. The *Guardian*'s science editor Tim Radford pointed out that the *Sea Empress*'s 50,000 tonnes of oil was a "drop in the ocean" compared with the annual 2.35 million tonnes of "chronic poisoning of the environment." Oil spills, said Radford, accounted for just two percent of the oil poured into the environment each year. He was skeptical about claims that the environment would take 10 years to recover and cited marine biologist Dr John Widdows, who said: "It all depends . . . . It's very complex: you can't predict exactly."

On *The Times'* editorial page of the same day, science editor Nigel Hawkes was even more forthright. The spill was "not catastrophic," and "within a few months . . . the evidence of the spill should largely have disappeared. Those of us who said this at the time of the Braer spill in the Shetlands three years ago held our breath, but only for a day or two. . . . The sea swallowed the oil as easily as many swallow the environmentalists' propaganda, trotted out shamelessly each time a tanker goes aground." Hawkes's article pointed to "bioremediation" as a remedy: the environment would mend itself. He concluded: "The paradox is that the environmentalists who criticize modern society for its dominion over nature are making the same error themselves, by implying our accidents are capable of destroying it." Hawkes notwithstanding, however, *The Times'* news coverage that day claimed: "If the *Sea Empress* breaks up, the environmental damage is likely to be worse than the Braer disaster . . . ."

Information about the *Sea Empress* came to the public mostly via the mass media. By raising questions of disputed fact and issues of policy, the media brought their audience into a debate and invited it to weigh arguments and make judgments. A general election in Britain was imminent, so how sections of the public judged the behavior of various players was of more than just academic interest. Given the conflicting voices in the media and the uncertain messages conveyed, just how were people supposed to make sense of it all? Just how "scientifically literate" would one have to be to sort out the contending viewpoints? And which of Durant's three definitions of "scientific literacy"—knowing a lot of science, knowing how science works, or knowing how science really works—would be most appropriate anyway?

## Knowing a Lot of Science

Take the "1001 things everyone should know about science" approach proposed by Hazen and Trefil, and one can imagine making up lists of essential facts about the various branches of science. For chemistry, the list might include the "solar system" model of the atom, some simple chemical formulas, and the flexible bonding capabilities of carbon. Physicists might want fundamental particles and conservation laws on their list. But what of the environment? Some of the items for inclusion might be that the Earth is a planet, and a very dynamic one at that, that the Sun is the source of energy for life, that burning fossil fuels produces carbon dioxide and that carbon dioxide is a "greenhouse gas," that there is a water cycle, and that plants and animals make up a continuous food chain. Simply looking at those few items points up one immediate problem: the environment covers many specialized disciplines, and the fundamentals will be drawn from many different lists. A sufficiently comprehensive list would turn even the experts into laypeople, because they would be unlikely to have the breadth of knowledge that they would suddenly find they needed.

In the case of the *Sea Empress*, the issues raised in just that one incident involved the chemistry and physics of oil/water/detergent interactions, marine engineering, hydrology, the toxic properties of oil and the vulnerability of various species to it, the various interactions of the local ecology, and the behavior and breeding of birds, fish, and seals, to name but a few, as well as risk assessment, tanker design, politics, and economics. Given that knowing a lot about all of these is a very tall order, the solution might be to prioritize; but exactly which of these many areas should the public know best if they are to participate in informed and rational debate?

Even where there might be a consensus on the area in which knowledge is most important, there can be disputes over the facts. How long, for example, do environments take to recover from the effects of acute pollution? According to a nature warden quoted on Channel 4 news, a unique submarine ecosystem was under dire threat. The message from Nigel Hawkes of *The Times*, however, was that ecologists had "cried wolf" too often and that nature was quite used to putting itself back together again. Was Tim Radford's statement in the *Guardian* that the 50,000 tonnes of oil spilled was a "drop in the ocean" compared to annual leakage rates likely to calm or alarm the reader? Radford clearly intended the former, but even an undisputed fact can have contentious implications.

## Knowing How Science Works

So if the facts of science are open to dispute, perhaps the processes of science can offer some insight into what we can know or believe about environmental science: there may be sense to be found in understanding how that science works.

Environmental change is, by and large, a chronic process, although it may be affected and effected by disasters such as the *Sea Empress* accident. But the media deal in the acute—who won last night's football match, or a celebrity's latest romance. As media scholar Anders Hansen has pointed out, in order to fit environmental issues into the media agenda, pressure groups grab attention by making claims tied in to well-organized publicity campaigns.[40] And if those claims can be dramatized by a daring and unexpected event, so much the better. So it was in the case of the derelict Shell oil platform, the Brent Spar. On May 1, 1995, campaigners from the environmental group Greenpeace boarded and occupied the rig 250 miles off the northeast coast of Scotland, claiming that Shell's plan to dispose of the rig by sinking it in the Atlantic Ocean would endanger marine life.

At the heart of the Greenpeace claim was that there was a total of 27,000 tonnes of lead, zinc, PCBs, oil residues, and radioactive scale on board the rig. Greenpeace expressed concern lest the deep-sea dumping of the Brent Spar set a precedent for some of the other rigs currently operating in the North Sea and claimed that the proposed disposal method flouted several international maritime agreements. Although Greenpeace activists left the oil rig after three weeks, a popular campaign against Shell forced the company to back down on June 21 and to look for alternative methods of disposal. Their decision left British Prime Minister John Major looking decidedly foolish, as he was busily nailing his colors firmly to the mast of deep-sea disposal just at the moment when Shell bowed to green pressure.

All along, Shell had argued that its decision to dispose of the rig intact in the Atlantic was environmentally sound and posed a much lower health and safety risk than any attempt to demolish the huge structure on land. In this, Shell was backed up by a sizable body of scientific opinion, including the findings of experts on the deep-sea environment in which Shell proposed to dump. Then, on September 5, Lord Melchett, the chair of Greenpeace UK, admitted that Greenpeace had made an error when sampling the tank allegedly containing the toxic oil residues and had overestimated the amount of potential pollutants. By this time, however, the decision not to dump at sea was irreversible. But the media had already begun to feel somewhat uneasy about the role they had played in the whole affair: they felt they had been duped by Greenpeace. On August 27, Channel 4 TV's senior news editor had told the Edinburgh National Television Festival: "On Brent Spar we were bounced. This matters—we all took great pains to represent Shell's side of the argument. By the time the broad-casters tried to intervene on the scientific analysis, the story had long since been spun far, far into Greenpeace's direction." Other media workers agreed, and the television channels, in particular, took steps not to appear partisan in their cov-erage of the Greenpeace campaign against French nuclear tests held later that year.

Given the media's role as the public eye on such an important problem as Brent Spar—there are over 400 oil rigs operating in the North Sea that will even-tually need decommissioning—the failure of the media to make a serious scien-tific analysis of the claims and counterclaims appears at first sight to be highly negligent. Indeed, Shell's case—which it claimed to have researched very thor-oughly before applying for permission to dump at sea—fell almost by default. The question here is: would understanding how science *works*, in terms of knowing how results are arrived at and how the consequences of those results are worked out, have led to an informed debate from which the "right decision" could have been reached?

Had the media described how both sides arrived at their results, the public would have been confronted by Shell explaining why the best option was to leave everything intact, and let a process of slow dilution by the sea get rid of any pollutants, while Greenpeace explained how it had worked out how much pollutant material was aboard the Brent Spar and why it mattered. Shell could not have answered Greenpeace's (flawed) pollutant estimates because, appar-ently, it had not worked out its own figures. Greenpeace could not have coun-tered Shell's deep-sea dumping claim because it had not had the opportunity to do the oceanographic modeling. On the face of it, the public would have been offered two apparently competent scientific cases, not necessarily con-flicting at the level of numbers derived from measurements and modeling, but from which very different conclusions on dumping policy were being drawn. In any case, by the time the flaw in Greenpeace's case had been revealed, the argument had moved on to more general issues of industrial secrecy (Shell promised to be less sneaky in the future) and whether it was right to consider each rig as an individual case or whether the cumulative effect was important. Anyway, the debate was not simply one of scientific rights and wrongs: dis-cussing the issue in *Nature*, Alison Abbott concluded: "Greenpeace's most suc-cessful argument is likely to be that no amount of scientific argument can offset a gut feeling that the sea should not be used as a toxic dumping ground under any circumstances."[41] Science, either as facts or as processes, was almost beside the point.

## Knowing How Science Really Works

Durant argues that the science that the public really needs to understand is "science-in-the-making," on which the scientific community has often yet to reach a consensus. The environment is very much a case in point here. Global warming, the destruction of polar ozone, the loss of rain forest—all these are areas in which data are hotly contested, and in which rival models attract both hyperbolic praise and withering scorn. Techniques of dealing with oil spill damage—chemical, physical, or biological—are similarly subject to considerable debate.

The wreck of the *Exxon Valdez* in Prince William Sound, Alaska, in 1989, was a disaster both in terms of its effect on the wildlife and beauty of the area and in public relations terms for Exxon. In September 1994, the company was forced to pay $5 billion in damages. This was less than had been expected, however, and Exxon shares actually rose that day. Its environmental stock had already risen with the publication in *Nature* of research by Exxon scientists that suggested that a relatively novel approach called bioremediation—encouraging nature to mend itself—could be the way forward for dealing with acute environmental damage.[42]

So would understanding how the Exxon scientists had arrived at this result, and appreciating how they had really drawn their conclusions, have enabled the public to work out whether bioremediation really was the answer or just an excuse by the oil company for doing next to nothing? Given Nigel Hawke's claim against the environmentalists in the case of the *Sea Empress*—that nature could look after itself despite humanity's worst excesses—could appreciating what this *Nature* paper revealed and how its results were obtained aid in understanding how the science of the environment really works?

First, the fact that the paper was published in *Nature*—a prestigious peer-reviewed journal—indicates that the work had been deemed both competent and important by an influential subset of the scientific community. Several other scientists would have assessed the research and have decided that it was worth bringing urgently and prominently to the attention of other scientists in a range of disciplines. The references to other research made in the paper indicate that the Exxon scientists were not out on a limb with their idea: it had precedents within science and was a step on a journey toward a solution that other scientists had already begun. However, this paper was only one more step in the generation of reliable knowledge. Once published in *Nature*, the research would have been subject to scrutiny by a broader group of scientists, who would then have decided whether or not the research could be taken further or applied more generally, such as in the case of the *Sea Empress*. The two years that had elapsed between the publication of this paper and the *Sea Empress* accident was a very short time on the scale of scientific research, and so by the time Hawkes drew attention to the possibility of bioremediation, he would have been talking about something that was still "science-in-the-making," and therefore still open to dispute, modification, and challenge. In his article, Hawkes did not give his readers any indication that bioremediation was the subject of ongoing and contentious research.

Nevertheless, given that the beaches of Wales were awash with crude oil and that an urgent solution was required, how much confidence could we have in bioremediation on the strength of the Exxon scientists' paper? The researchers themselves acknowledged the very complicated circumstances in which their idea would have to work. They had looked at three different types of beach in Prince William Sound—a "low energy" beach, hardly disturbed by wave action and contaminated only with surface oil; a similar beach with both surface and subsurface contamination; and a deeply contaminated "high energy" beach that was subjected to heavy pounding by the sea. They had treated each of these beaches with various types of fertilizers containing nitrogen and phosphorus, to supply naturally occurring or deliberately introduced oil-degrading bacteria with the nutrients needed for maximum growth. The details of the *Nature* paper made it clear—and the accompanying "News and Views" article "Bioremediation comes of age"[43] even clearer—that the work was carried out under specific conditions of beach type and wave action that would limit the applicability of the results of other oil pollution incidents. That had to be understood if Nigel Hawkes's claim for the efficacy of bioremediation was to be weighed carefully in the case of the *Sea Empress* accident.

What about the time it would take for bioremediation to work? Time was important in the case of the *Sea Empress* in view of the imminent arrival of breeding seabirds on the nearby nature reserves. In the paper describing the pollution of Prince William Sound by the wreck of the *Exxon Valdez*, the *rate* at which the level of oil pollution decreased was fitted to a multicomponent exponential equation, normally not the stuff of science popularization. So there were many parameters, which might change from situation to situation, that affected this rate. Furthermore, the overall time it would take to clean a beach by bioremediation would also depend on how dirty it was in the first place. How, then, could one be sure that the Welsh beaches affected by the *Sea Empress* would be cleared of oil in time for the birds to arrive and breed in safety?

The authors of the *Nature* paper also discussed what chemical fraction of the oil they should monitor as a measure of the degree of pollution. This was vitally important, since oil is not a simple chemical but a complex mixture. In different oils—and thus in different cases of pollution—the proportions of the various fractions of this complex mixture vary: one might be monitoring an important fraction in one instance, but, in another, this fraction would turn out to be only a minor constituent, and thus not a good indicator of how rapidly the oil was being cleaned up. And to make things even more complicated, other variables in the Exxon Valdez study included when to apply the fertilizer, how to control for sources of nitrogen that were already present in the beach, and how deep to dig for samples. Overall, the paper showed that the scientists had a number of decisions to make about how they should proceed, any of which could have affected the applicability of their results to other situations.

There are also wider considerations in problems like an oil-soaked beach. Among those concerned with dealing with oil pollution are to be found advocates of mechanical and chemical cleanups. Still others feel that doing absolutely nothing—leaving nature, unassisted, to deal with the problem—is the best way forward. Outside of the scientific and technical community, issues of public concern include whether it is right to introduce gas-guzzling bacteria into oil-polluted environments where none have existed before.

Thus, the process of understanding how the science of bioremediation really worked in the case of Prince William Sound, and whether it really was a method that could be applied to other instances of oil pollution, was very time-consuming and required background scientific knowledge of considerable breadth and depth, as well as a grasp of ecological ethics. All of this seems to complete the circle back to where we started: at the definition of scientific literacy as the need to know a lot of science. And understanding that science-in-the-making is not yet reliable knowledge, and that at this stage no scientist or activist can claim "the right answer," leaves the public with little to rely on other than their instincts or common sense. Each of Durant's definitions takes us only part of the way toward scientific literacy: the whole story is clearly more complicated—in theory as well as in practice.

## IN CONCLUSION

The examples described in this chapter emphasize that as science has become more and more enmeshed in everyday life, so public science has become more and more complex in nature and import. Faraday and his contemporaries blazed a trail for science communication in the mid-19th century. By today's sophisticated, multimedia standards, unashamedly didactic expositions like Faraday's on the candle are considered old-fashioned and suitable, if at all, only for children as an adjunct to their education. However, while "knowledge" in terms of facts does not constitute understanding, it is still something that many people enjoy acquiring for its own sake: that there remains widespread curiosity about how ordinary things work was demonstrated in 1989 by the success of the book *The Pencil* by engineering professor Henry Petroski, who put the humble pencil in place of Faraday's candle and took an extensive tour of its history, chemistry, engineering, design, and manufacture.[44] The mass media demand is for entertainment as well as education; so science adventure stories such as $C_{60}$ fit the bill better than Faraday's moral homilies. But most science is not as straightforward and as reliable as the chemistry of candles or pencils, or even of exotic forms of carbon. Such simple subjects are rarely the stuff of headlines or political debates.

However, complexity does not, in itself, constitute a barrier to widespread media coverage of science. With its high stakes and relative paucity of data, cosmology is a notoriously difficult and emotionally charged science, even within the scientific community. But, as cosmologists ever since Einstein have demonstrated, laypeople do not have to understand science in any depth in order to enjoy cosmology or gain from it some insight into their world. Cosmologists are still a relatively new species, but the questions they ask about origins are the same ones that people have been asking in various ways for centuries. It is hardly surprising, then, that cosmology and the public mood resonate or that, therefore, cosmology is the stuff of the mass media.

But cosmology does not impact directly on daily life. When hard choices are to be made and practical action taken, "objective facts" may seem the best basis for decision making. Will the environment take care of itself, with maybe just a little help from its friends, no matter what we dump into it? Should redundant oil rigs be sunk in the deep ocean? Should single-skinned tankers be banned from coastal waters? Objective facts in these debates are hard to come by, however. So it is not easy to be optimistic about the prospects of the public really being able to come to grips with scientific debates about the environment, at least while the debates remain only scientific. In the arena of the environment, at least, it would appear that battles are to be won and lost more on emotion than on a disinterested and objective scrutiny of scientific facts. As we saw in Chapter 4, sociologist Friedhelm Neidhardt argues that this is an inevitable consequence of conducting such battles through the mass media[45]; and when the facts are themselves contentious and contradictory, they make only poor weapons in the hands of those who would argue for a more objective debate. The dream of educators like John

Dewey that problems in a democratic society could be resolved by rational discussion, provided one first taught people to think "scientifically," seems just that—a dream with little chance of being realized.

Each of the concepts of scientific literacy that we have looked at has its problems: of being over-simplistic, of being incapable of distinguishing between rival scientific claims, and of requiring too much time and scientific knowledge to be a realistic option for most members of the general public. The path to "scientific literacy" seems to be endlessly circular. If there is room for a more upbeat position, however, then maybe it is in the way the often much maligned media are starting to treat environmental issues. If the coverage of the many points of view on the *Sea Empress* becomes the norm for dealing with the "disaster scenario," and if the self-criticism the television companies indulged in over the Brent Spar oil rig is taken to heart, then readers and viewers should be presented with a more sophisticated and more complete picture. The circle might turn out to be an ascending spiral after all.

One thing emerges clearly: much of the science that the public needs to know about, and most of what the media present, is new science, science-in-the-making, and is therefore likely to be both provisional and controversial. It is very unlike the tried-and-tested science that is found in textbooks, and about which the public are quizzed in the questionnaires of survey researchers. Science-in-the-making put strains on everyone involved in the process of public understanding of science: on scientists, in knowing what to claim; on journalists, in assessing what is reliable and signficant; and on the public and their representatives, in matching the new facts and ideas to what they already know and how they already live, and in deciding what to do. In cases like these, when scientists themselves are struggling to understand science, journalists and the public face extra challenges in understanding the scientists. Such situations are particularly problematic when risks to people are involved—either to individuals or to the public at large. Then, working out the "right response" cannot be achieved exclusively from within the realm of science. We look at that next.

**Chapter 7**

# An ABC of Risk—Apples, Beef, and Comets

**After a century in which science and technology have changed societies and** invaded personal lives to a far greater extent than ever before, we are learning to live with the risks as well as the benefits of science. There are very few technologies that do not present some hazard, however small, to the public. Those technologies which endure are those whose benefits outweigh the hazards—where we have decided that the risks are worth the advantages the technology brings. Sometimes a new risk arises, because of new technology or because a tried-and-tested process has suddenly found a new and unexpected way of going wrong. Sometimes risks that have gone unnoticed are made visible by new research, and the lead pipe and white asbestos are thrown away: in this sense, science "creates" the risk by exposing it, just as it often creates or exposes riskiness in nature or in new technologies.

Scientists have long held the responsibility for understanding the dangers of life in a technological society. Until they come to the attention of the public, potential hazards are completely in the technologists' hands. Scientists and engineers choose who will run the risk, as well as how great the risk will be, when they choose their model or design. When the hazard becomes public, all manner of other professionals rush on to the scene—managers, financiers, politicians, public relations people—but the scientists they cite are seen as the ultimate arbiters of truth, as purveyors of authoritative knowledge, and as the last word in the risk society. According to science policy scholar Sheila Jasanoff, expertise is all-important in modern life:

Without authoritative, expert institutions, we could not be reasonably sure that the air is safe to breathe, that aeroplanes will take off and land safely, that new medical treatments will not unexpectedly kill patients, that chlorofluorocarbons will not eat away the earth's ozone shield— and . . . that the food we buy is safe to eat. Lives lacking such assurance would be impossibly difficult to cope with, both pragmatically and psychologically.[1]

So when the public is worried about the food it eats or the air that it breathes, who better, then, to advise and reassure than the scientist? Science is widely believed to offer the objective truth: a scientist's words therefore carry great weight in times of uncertainty. But, as we saw from the environmental pollution examples in Chapter 6, when different groups with different agendas each proffer a different scientific truth, claims of scientific impartiality and authority seem very feeble indeed.[2] No wonder, then, that the public loses patience with science when one scientist says one thing and another scientist advises the opposite.

The psychology of risk is complex. The enterprises of science create and define risks, and the public perceives risks; but the definition and the perception can be very different.[3] Thus, expert and lay assessments of risk are sometimes very different: for example, the public tends to overestimate visible or sensational risks, such as road accidents or murder. The acceptability of a risk is not connected in any straightforward way to the degree of risk: for example, people are prepared to accept relatively high risks from activities they choose for themselves, such as skiing, smoking, or driving their own car, but will object to risks that are technically much smaller if they are outside their control, such as the possibility that power cables might cause cancer or that a terrorist might put a bomb on an airplane. The public tend to be less happy about a risk the more people it might affect, and are often willing to accept a hazard to themselves that they would not support were it proposed for others.[4]

But the world is not in a state of constant mass panic: people are more likely to see themselves as immune to danger. Most people claim to be better than average drivers and think that they will live to over 80. Obviously, these claims are psychologically sustaining on a personal level; but they are also rational possibilities, within the bounds of even the most technical risk statements, which say something about what might be in general, and nothing about what is in any individual case.

Risk assessment is a mathematical expression of probabilities and severities, not an immutable explanation of the scientific situation. Further, like all forms of statistical analysis, risk assessment applies to

the average, not the particular. For example, the likelihood of a nuclear
power plant accident occurring is one in every 10 000 reactor years—
an industry average but not an analysis of any single plant.[5]

So the public relations consultant who reassures you about the safety of nuclear
power and the activist who tells you that your local power plant is a disaster waiting to happen have both interpreted the risk statement correctly, if incompletely.
In this sense, risk statements are a godsend to anyone with an agenda—and much
less use to anyone who simply wants to know how to live a safe life.

Yet the differences between scientific statements of risk and the public's perception of and reaction to risks are often dismissed or even condemned as evidence
of public irrationality or ignorance by technological decision makers, who conclude
that any resistance to or fear of technological risk is simply a problem of information: people could be made to react as rationally as scientists if only they knew the
facts.[6] Risk communication scholar Susanna Hornig notes that risk stories in the
mass media "are typically judged on how accurately they reflect the scientific point
of view and how well they contribute to public education designed to eradicate
wrong thinking."[7] As the following case studies show, information does seem to be
the central weapon mobilized, largely through the mass media, to diffuse the anxiety
and opposition generated when citizens feel at risk from science. But is information
enough? This chapter examines the reactions to three risks: the risk of cancer from
eating apples sprayed with Alar, the risk of Creutzfeldt–Jakob disease from eating
beef infected with the "mad cow disease" bovine spongiform encephalopathy
(BSE), and the risk of global destruction from a collision between the Earth and a
comet. We look at how scientists, scientific institutions, and the media responded to
these fears and consider how successful these responses were in negotiating the
complex dynamics of the relationship between science and the public.

## APPLES AND ALAR

Apples have a special place in American hearts and stomachs. Apples represent all
that is wholesome; in pies they define Americanness; a big apple symbolizes New
York. What is more, fruit is a "health" food: low in fat and high in vitamins, apples are, metaphorically and literally, "good." It was a particular shock, then, for
Americans to learn, in 1989, that apples could give their children cancer.[8]

The alleged carcinogen was not in the apples themselves: it was sprayed
onto them in a ripening agent called Alar, which contains daminozide. Both
daminozide and a derivative, dimethylhydrazine (UDMH), had been found in

apple juice and other prepared foods containing apples. Alar had been approved for use on apples in 1968 and was manufactured for the apple industry by the company Uniroyal Chemical. But within a few years, studies had indicated that daminozide and UDMH were associated with cancer in mice, and by 1984 the Environmental Protection Agency (EPA), the U.S. government department responsible for agricultural safety, was sufficiently persuaded of the potential problem to begin a review of the suitability of daminozide for food use. After several years of internal debate, the EPA recommended that Alar could still be used on food, but at slightly lower levels and on condition that Uniroyal Chemical submit new data on the toxicology of daminozide and its derivatives.

In 1989, in light of the new data, the EPA estimated that the risk to adults of cancer from a lifetime of daminozide in the diet was five additional cancer cases per 100,000 people—a much greater risk than the EPA usually allowed. However, after reviewing the data and recalculating the estimate, the EPA found the risk to fall within reasonable limits—one extra cancer case per million people. But for children, because of their different diet, the risk was nine extra cases of cancer per million people, and so the EPA proposed canceling the food use of daminozide.

At the same time as the EPA was calculating these estimates, the National Resources Defense Council (NRDC), a well-known nongovernmental environmental pressure group, published its own analysis of the EPA data. The NRDC concluded that exposure to daminozide would cause one additional case of cancer for every 4200 children aged 6 and under. The NRDC publication *Intolerable Risk: Pesticides in Our Children's Food* charged that the government's regulation of the use of chemicals in agriculture was irresponsible and that safe levels had been calculated for adults and were inappropriate for children. Public relations consultants were hired to handle *Intolerable Risk*, and they arranged exclusive rights to the story for the CBS news program "60 Minutes." When the program was broadcast in February 1989, to an audience of 40 million people, it caused what environmental law researcher Kerry Rodgers describes as "national pandemonium." *Intolerable Risk* was launched at press conferences around the country the next day, and actress Meryl Streep appeared as a spokesperson for the NRDC on talk shows and in a television commercial. Schools stopped serving apples, supermarkets stopped selling apple products, and the U.S. Congress started to debate legislation to override the EPA. By one account, a concerned parent asked the police to stop her daughter's school bus and remove the apple from her lunch box.[9]

The apple industry began a counterattack, orchestrated by another public relations firm, and both sides of the argument began to appear in the press. An edition of "60 Minutes" broadcast in March 1989 included appearances from several critics of the NRDC's stance. Elizabeth Whelan, of the American Council on Science and Health, claimed that the evidence linking Alar and cancer was inconclusive and that overreacting against pesticides could lead to food shortages. Biochemist Bruce Ames, director of the National Institute of Environmental Sciences at the University of California at Berkeley, said that there was more risk from cancer from driving an extra mile to buy organic fruit. The apple industry announced that anyway only a small proportion of apples were sprayed with Alar; Uniroyal Chemical declined to comment.

The story had petered out in the media by the autumn of 1989. Uniroyal Chemical had stopped selling Alar in June; the apple industry had lost millions of dollars, and supermarkets were grilled for some time by customers about the chemicals used in the production of fruit and vegetables. The scientific controversy continues, and conflicting results are still produced from studies of the carcinogenicity of Alar. But what did the Alar scare mean to the various groups involved? According to Rodgers, various interest groups adopted Alar as a symbol of their own particular political, social, or scientific agenda: Alar was mobilized as a weapon in a battle that went far beyond the question the public was asking—whether or not apples were safe to eat. For the environmental and consumer lobby, spearheaded by the NRDC, Alar represented the government's failure to protect its citizens, and its loyalty to corporate rather than consumer interests; the public and media outcry reflected a lack of public confidence in the government's ability or willingness to reduce risks. According to an NRDC spokesperson, the media, having simplified the story for popular consumption, had lost sight of the broader issues when they focused on apples.

The significance of the Alar scare for the industries affected was, according to Rodgers, that it exemplified the improper use of science by activists and the irresponsible behavior of the media. The apple industry enlisted scientists to counter the apparently scientific claims of the activists and charged the media with irresponsible and sensational reporting. Uniroyal Chemical lamented the misinformedness of the environmentalists, and a spokesperson for an agrochemicals trade association contrasted proper scientific decision-making with "emotion-driven stampedes like the . . . Alar experience." For the scientific community, the Alar scare showed the extent to which activists were prepared to frighten the public in order to draw attention to their political and ideological agenda. Editorials in the American Association for the Advancement of Science's influential journal *Science* criticized the NRDC for oversimplifying scientific information in order to exploit public fears and to attract new members, and charged the media with hyping a story with little concern for its credibility.

Every group, then, had something to say about the role of the media in the Alar scare. The extensive coverage was thought by many people to have been overly alarmist and irresponsible—indeed, some went so far as to claim that it was the media, rather than Alar itself, that were at the heart of the problem. Even journalists criticized the coverage: according to Marla Cone of the *Los Angeles Times*, for example, the reporting of Alar was "outrageous...completely alarmist." Environmental journalists, worried that they had unnecessarily unleashed a monster, started to question their own professional philosophies and ethics. The coverage is even said to have prompted nine states to make the libel and slander of fruits and vegetables offenses in law. In light of these reactions, professor of journalism Sharon Friedman and her colleagues undertook an analysis of the risk information in the media coverage of Alar, to see whether or not the media could be held responsible for the apple panic. According to Friedman et al.:

> Reviewing media coverage of Alar also brings into focus the important question of whether the media are performing responsibly when dealing with risk issues. This is very important since so much is at stake for all parties concerned: industries being regulated, government regulators, watchdog groups involved, and members of the public who might be scared needlessly or, conversely, not alerted sufficiently to potential dangers.

So was the coverage alarmist? The study by Friedman et al. of 297 articles from 13 newspapers indicates that, overall, it was not. For example, the most quoted source in almost 20 percent of the coverage was the apple industry, which sought to reassure consumers; in contrast, the NRDC, which sought to warn of the dangers of Alar, was the most quoted source in less than 7 percent of articles. Adding together all the groups opposing the use of Alar still finds the anti-Alar voice most prominent in less than 18 percent of articles. Most prominent overall were the government agencies, in 42 percent of articles, and the apple lobby, in 34 percent of articles.

The debate about Alar between the interest groups involved was a debate about the size of a risk—the risk of getting cancer as a result of eating apples—but, according to Friedman et al., the coverage of information about this risk, and about the different risk estimates proposed by the different parties, was extremely limited. Only 16 percent of articles included the size of the risk. Only 11 percent mentioned the EPA figure, and seven percent mentioned the NRDC figure. Only three percent of articles mentioned both figures, so only these articles actually gave the conflicting data that had caused the debate in the first place. Some articles picked out an NRDC statement that 4800 children would get cancer from apples; in others, risk information was barely mentioned.

Only four percent of articles mentioned any other risk that readers could use as a yardstick. In one story, Bruce Ames provided examples of foods which naturally contained much higher levels of carcinogens than apples treated with Alar. In another, a senator claimed that smoking, driving, and drugs kill people, but apples do not. Only 15 percent of articles mentioned that other foods had risks attached.

Almost 85 percent of articles included no information on how much Alar was needed to cause cancer. Those that did mention a dose gave the figure in parts per million, which most people would have found very difficult to relate to their own consumption of apples. The apple industry said that the dose that caused cancer in rats was the equivalent of feeding a child 28,000 pounds of apples. Cancer was mentioned in 75 percent of the articles, and 48 percent mentioned that children were particularly at risk. What that risk was, though, was not specified.

Four of the 13 newspapers analyzed in the study by Friedman *et al.* were from regions such as New York, California, and Washington where apple growing is a major industry. These newspapers dealt more thoroughly with the risk information in their stories, providing around half of the total number of articles in the study that mentioned dosage, conflicts over data, and other food risks, and 61 percent of the articles that discussed the link between Alar and cancer.

Friedman and her colleagues concluded from their study that on the basis of the 13 newspapers they analyzed, claims that the coverage of Alar was responsible for the ensuing panic were unjustified. The newspapers certainly were effective at alerting people to the issue, but neither headlines nor lead paragraphs were alarmist. This raises the problem of how, then, to account for the fact that panic did indeed ensue: Friedman et al. suggest that the sheer volume of coverage, in so many media, could have been responsible.

The newspapers also gave comprehensive attention to the different interest groups involved: in this regard, the reporting was balanced. Newspapers from apple-growing regions in particular provided thorough in-depth coverage of the Alar story. However, the amount of information about the risks was very small: most stories did not mention the use and interpretation of the risk estimates by the different groups involved in the controversy. According to Friedman et al., the coverage did little to empower citizens at a difficult time:

> Reporters appear to have chosen to cover the controversy itself rather
> than interpretations of the science that caused the controversy in the
> first place. Such coverage does nothing to help readers understand the

perspectives of the various parties, nor does it give them enough information so that they can make various decisions about the issue, such as whether they should be concerned about it, try to get more information, or stop feeding their children apples.

## BEEF AND BSE

The relationship between the British and their beef has been as close as the relationship between Americans and their apples. But as in the United States, so in Britain: a perceived risk to people's health overrode their patriotism and changed their eating habits as "mad cow disease" erupted in British beef herds and on the front pages of the newspapers.[10]

In England in the spring of 1985, a cow died of an unknown degenerative brain disease. Soon other cases were identified, and in 1986, the disease was named bovine spongiform encephalopathy, or BSE. It quickly became clear that BSE was similar to diseases already known in other species, including scrapie in sheep and Creutzfeldt–Jakob disease (CJD) in people. In April 1988, the British Ministry of Agriculture, Fisheries and Food (MAFF) commissioned an inquiry, chaired by Sir Richard Southwood, into the epidemic and its implications for human health. BSE soon became a notifiable disease, and a regulation was imposed requiring all infected cattle to be destroyed.

By now, scientists had come to suspect that the source of the infection was cattle feed containing offal from sheep infected with scrapie. In March 1989, the Southwood inquiry reported that increasing use of feeds derived from sheep offal, together with changed methods of feed production, was the most likely cause of the problem; and the government banned the use of offal in cattle feed. The inquiry advised that it was "most unlikely" that BSE would have implications for human health; and, assuming no vertical transmission (from mother to calf) or horizontal transmission (from cow to cow), it concluded that the disease would become extinct within a decade. It did note, however, that if its risk assessment were wrong, the implications for public health would be serious. So Southwood also called for further research and careful monitoring.[11]

Throughout the late 1980s, BSE was mentioned only occasionally in the media. The coverage was of the publication of new reports and the implementation of new safeguards, and it was confined to the specialist scientific press and the science pages of the quality daily press. The stories were, after all, stories about science. Toward the end of 1989, however, media interest began to grow. Scientists had produced some new ideas about the disease and its causal agent;

but, more importantly, Germany banned imports of British beef because of the risk it posed to public health. In an outburst of indignation and wounded national pride, BSE was suddenly a story fit for the tabloids.

Food safety was high on the public agenda at this time: several common foods—cheese, yogurt, and eggs, for example—had recently come under scrutiny. Early in 1990, the *Guardian* newspaper pointed to growing public unease about meat eating. Grouping BSE with a number of other public health issues in the news, it suggested that the introduction of cost-saving production techniques was compromising the safety of meat and undermining public confidence. It reported that a government survey revealed "a fairly steep fall in the consumption of meat, and particularly red meat."[12]

This was the climate in which, in May 1990, *Today* newspaper announced "Mad cows' disease kills cat."[13] Urgent tests were being carried out to see how the cat had contracted the disease, and whether the disease could be passed to other species. Five days later, *Today* announced that it had found some unpublished research that offered: "Scientific proof: mad cow link to humans."[14] BSE was now a story about people. But public suspicions had already been roused, and some agencies had already taken action: beef had been taken off the menu in schools and in nursing homes for the elderly and, based on "general medical opinions from our medical Fellows," was no longer being served at high table in Magdalene College Cambridge.[15]

Dead cows were one thing, but the collapse of the beef market was quite another. John Gummer, Minister for Agriculture, Fisheries and Food, leaped into the fray, determined to allay public fears. Relying on the advice of the Southwood inquiry, Gummer claimed that British beef was safe. He said: "My wife eats beef; my children eat beef; and I eat beef: that is everyone's absolute protection."[15] Gummer's deputy, David Maclean, was pictured in the press alongside a sausage. He said:

> The thing that annoys me is that we have published all the facts about BSE . . . . When you get into a situation where people don't want to reassurance and want to scare themselves to death there's nothing you can do about it. Commonsense and science seem to have gone out of the window. The message to the public today is the same as it was yesterday, last month and last year. British beef is safe.[16]

Later, Maclean said:

> If there are some people who do not want to believe that it is safe, God help them—let them wallow in their own pathetic little panic. But let them keep their mouths shut and not scare the vast majority of us who have common sense.[17]

Gummer's remit at the Ministry of Agriculture, Fisheries and Food was twofold: he had responsibilities both to the farming and fishing industries and to consumers. This, as the cartoonists looking for cow jokes were quick to realize, put him on the horns of a dilemma: could Gummer, in the case of BSE, serve both communities? Opinion polls showed that it was widely believed that the Ministry had the interests of farmers rather than consumers at heart. When Gummer fed his daughter Cordelia a beef burger in front of press cameras, the satirical magazine *Private Eye*'s cover photograph expressed what many people felt: in superimposed thought bubbles, a photographer says: "The public won't swallow this," and Cordelia adds: "Neither will I."[18]

MAFF's attempts to reassure the public looked less and less credible. Back in November 1988, the Ministry's chief veterinary officer Keith Meldrum had announced that there was no risk of BSE transmission through milk, but a month later the sale of milk from cows with BSE was banned. In February 1990, plans were announced to investigate transmission via milk. In January 1990, Meldrum stated that BSE was not a risk to people. A few days later, a research program to investigate links between cattle and human encephalopathies was announced, and Meldrum claimed that it was "too early" to say that there was no risk for people from BSE.[19]

Farmers were disappointed with their Ministry. The president of the National Farmers' Union gave his minister a public dressing down:

> People recognize when there are voids in information. At the moment there is little confidence in official statements, and that is why there is mayhem . . . there is fear of the unknown.[20]

But the farmers were, for the most part, keeping quiet. Unlike the government, they had not waited for scientific reports before they took action on the farms. As repositories of generations of wisdom about animal husbandry, the farmers had swiftly spotted the connection between scrapie and BSE and had long since stopped feeding cattle on feed derived from sheep.

After the tabloids' reactions to the dead cat, a *New Scientist* editorial expressed the weary disdain of many scientists at what they saw as yet another unjustified lapse of faith in and understanding of science:

> One cat dies, "A Scientist" speaks, six million cows are supposed to hit the knacker's yard and you can't even turn the bovine bodies into feline fodder. Or, to put it another way, hysteria rules . . . the media are at it again . . . .[21]

A number of scientists spoke out. There was no evidence that people could catch BSE, they said, but there was no evidence that people could not catch it. The scientists were disturbed that the government was claiming scientific endorsement for the statement "British beef is safe" on the grounds that there was no evidence to say that it was not; there is no justification, the scientists said, for calling a statement "scientific" when there is no evidence for or against it and no way of testing it.[22]

Of the independent scientists who spoke to the press, the one who attracted the most controversy was microbiologist Professor Richard Lacey of Leeds University. Lacey had achieved some notoriety in Britain for exposing what he sees as dangerous practices of modern food production and has experienced considerable peer pressure to keep quiet (now retired, he writes science fiction thrillers as a way of informing the public about food and the environment). In the case of BSE, said an editorial in *Nature* in 1990, Lacey "went too far."[23] He had said that since the incubation periods for spongiform encephalopathies were of the order of tens of years, no-one would know if people had caught BSE for some time, and then we might find that an entire generation had been wiped out. The only logical approach for the human population in the United Kingdom was to avoid all beef products. Lacey pointed out that for doctors—that is, those experts who are primarily concerned with the well-being of people—it is normal practice to assume the worst and to act accordingly. Those with other priorities could be expected to take a different course of action.[24] Other scientists responded that Lacey's statement was just as irresponsible as Gummer's "beef is safe": there was no scientific evidence for either of them, and the public should be protected from scaremongering just as much as from false reassurances.

The meat industry stance was unambiguous. The Meat and Livestock Commission, which promotes the British meat industry, placed full-page advertisements in the national press that stated simply "beef is safe." In May 1990, the managing director of a chain of high-street butchers offered a strategy for calming the market: he wanted to "move the discussion of a complex scientific subject from the front pages of the tabloids to the back pages of the quality press . . . ." He added that as a contingency measure he had placed a very large order for French turkeys.[25]

What sense did the public make of these conflicting agendas? Letters to the newspapers showed the public to be impatient with and mistrustful of bland government reassurances, with people preferring to decide for themselves how best to react. As the author of a letter to a newspaper remarked: "I, for one, am not prepared to be the subject of an experiment, the unpleasant outcome of which may not emerge until some five years hence. Fortunately we have the freedom to choose to consume or boycott beef."[26] Polls conducted at the height of the episode—the third week of May 1990—showed that a majority believed that MAFF could not be trusted to deal effectively with issues such as BSE, or to tell the truth about it. Ninety-five percent had heard of BSE, and two out of three expressed concern. One in seven believed that all beef was infected, and one in seven that it had spread to household pets. Twenty-eight percent were eating less beef, and 10 percent less of all types of meat.[27] Another poll showed that one in 10 British people was more afraid of BSE than of AIDS and that 60 percent thought that the government was not doing enough to stop BSE from spreading to people.[28] Supermarkets said that beef sales had dropped by 50 percent but that sales of other meat had risen to compensate. The Meat and Livestock Commission said that sales had dropped by 25 percent, and claimed an "amazing 80 percent degree of confidence among shoppers."[29] A month later, sales were still down by 20 percent.

Like Alar, which made little impact on the British media, the mad cow episode of 1989–90 did not cross the Atlantic. Indeed, a few years later, during the next surge of concern in Britain over BSE, a group of risk communication students at Cornell University thought that the BSE fiasco was so daft that their tutor must have invented it as a joke. But when the British Health Secretary announced to the House of Commons in March 1996 that a cluster of cases of Creutzfeldt–Jakob disease in young people was "most likely . . . linked to exposure to BSE," mad cow disease made headlines around the world.[30] The chair of the British government's advisory committee on spongiform encephalopathy suggested a possible epidemic of half a million future cases of CJD.

Media coverage soared.[31] Distinguished science writer Richard Rhodes wrote a number of articles for the U.S. press and a book that called BSE the "new Black Death" and "a terrifying new plague."[32] British newspapers contained half as many articles on BSE in two months as they had in the previous 12 years.[33] The newspapers reported an all-time low in consumer confidence in beef,[34] and the mass media generally had a new angle on BSE: now they could feature real human victims. Friends and relatives of CJD victims told poignant stories of the progress of the disease, which causes paralysis, loss of speech and memory, and personality changes. Since the first BSE scare in 1989, it had consistently been stressed that the mysterious agent of BSE would be concentrated in the nervous tissue of cows, so that more expensive cuts such as steaks would be less likely to be affected than the cheaper pies and burgers, which might contain offal and nervous tissue. The father of a young single mother who had died of CJD sat in his orphaned grandchild's nursery and told BBC television news how his daughter had struggled to feed her child well but had been able to afford only the very cheapest meat for herself. Thus, CJD became not just a tragic and horrifying disease, but also a disease that victimized the poor.

An article in the *New Yorker* later in the year recounted the last months of Maurice Callaghan, whose death from CJD was the first to be recorded by a coroner as related to BSE. In the early stages of the disease, Callaghan had failed a psychiatric test because he could not name the Prime Minister of the United Kingdom, but, as his brother pointed out, "that may be because it's John Major." Major's government was struggling through its final months after a long, exhausting term of office and had become a nonentity in more than just this aspect of public life. Government agencies seemed unable to offer official guidance to consumers, although officials and politicians insisted that their families would still be eating beef.[35] Jasanoff perceived a "feeling of abandonment" that she described as "civic dislocation": "an unprecedented breakdown of communication between British citizens and their public institutions in the aftermath of the announcement of 20 March."

> The phenomenon that I call civic dislocation expressed itself as a mismatch between what governmental institutions were supposed to do for the public and what they did in reality. In the dislocated state, trust in government vanished and people looked to other institutions for information and advice to restore their security. It was as if the gears of democracy had spun loose, causing citizens, at least temporarily, to disengage from the state.[36]

Deprived of expert advice, consumers and retailers resorted to common sense—something that many people had long since felt was their only reliable resource. There are alternatives to beef, even in the British diet; and there are, in the international market, alternatives to British beef. Britons worried about beef had long since switched to other meats or had given up meat altogether. Scrapie was out of the food chain; farmers were only selling beef cows that were too young to have eaten the banned infected feed. Enterprising "organic" farmers introduced potential consumers to their prey in the field and allowed them to inspect its living conditions and its feed. And once again, the fear quickly subsided: in 1997, after years of trumpeting that only foreign beef was sold in their burger restaurants, even the fast-food chains announced that British beef was back on the menu. According to Health Minister Stephen Dorrell, the people dying of CJD had been exposed to BSE before the 1989 restrictions had been imposed. In July 1997, the Meat and Livestock Commission announced a 0.6 percent rise in sales, which they described as "a huge vote of confidence"; and the National Farmers' Union said it was "immensely grateful to consumers."[37]

When the Labour government was elected in 1997, one of its manifesto pledges—a suggestion from the independent Consumers' Association—was to separate government involvement in agriculture from government responsibility for food and consumers. The new food agency, entirely separate from MAFF, was announced in January 1998. It will certainly have beef on the agenda, for, by the end of 1997, new research had suggested that bones might harbour the agents that cause CJD. The response of the British government was to ban the sale of beef on the bone. Demoralized farmers responded by blockading ports and tipping beef imports into the sea; consumers stocked up with ox-tails and T-bone steaks ahead of the ban.

BSE in Britain, like Alar in the United States, provides an example of how public scientific debates can be comprised of combinations of institutions and institutional agendas and of how these different agendas may inform radically different public representations of the same issue. In the case of BSE, government spokespersons represented science as providing certainty—despite many caveats from the scientists themselves—in order to reassure the public, as in the early statements of the government's chief veterinary officer Keith Meldrum. Science was also represented as carefully weighing up the inconclusive evidence to make a risk assessment. The neuropathologist Richard Kimberlin, for example, was quoted in the *Independent* thus: "On the one hand, it would be wrong to assume that there is no risk whatsoever. On the other, any preventative action has to be taken knowing that it may be quite unnecessary."[38] A third representation of science was one of knowing what it did not know, and being honest about it. As Richard Lacey had said: "The only logical approach for the human population in the UK is to avoid all beef products."[39]

The BSE case also provides ample evidence that a key variable at stake in the interplay of different agendas is public confidence in the institutions of science and government. According to Jasanoff, these institutions

> require a set of background conditions in order to carry out their tasks in ways that merit public confidence: these include a widely shared and unambiguous problem definition, relative certainty about the relevant "objective facts," clearly identifiable expert knowledge about these facts, a reasonable convergence of societal values, and a more or less bounded space for the articulation of views and conduct of deliberations. When these conditions are present, day-to-day engagement between experts and citizens may indeed be superfluous. A discreet, well-insulated process, founded on expert judgment, may be quite capable of producing decisions that are balanced, persuasive, efficient and, most important of all, right. . . . In the BSE case, as in a growing

number of environmental conflicts in advanced industrial societies, these conditions simply did not exist.

In Jasanoff's analysis:

> Given the pervasive uncertainties surrounding BSE, the normal distance between citizens and experts was greatly reduced. The scientifically uninformed public was almost as well positioned as the experts to make sensible decisions about how to avoid the ill-defined and poorly characterized risk of BSE—for example, by adopting various dietary restrictions. Yet, the UK's characteristically insulated and confidential decision-making process excluded wide public involvement until the government's disavowal of any risk to human beings was shown to be unfounded. Ironically, the British public learned only after the fact that government experts . . . had been taking the kind of common-sense precautions (not feeding beef to grandchildren) that they too might have wished to take had they been better informed about the uncertainties of BSE transmission.[40]

# COMETS IN COLLISION

Some people might wish to hold scientists and their allies responsible for poisoning our food, but it would be difficult for even the most ardent anti-science activist to blame anyone if a comet smashed civilization to smithereens—assuming, that is, that anyone survived. From the perspective of the public, a "cosmic catastrophe" such as a comet impact on Earth is the kind of potential hazard that does not exist until scientists decide to create the risk by making their knowledge public.

Astronomers have known for many years that the space we inhabit is a bumpy place. For the first 750 million years after the Sun began to shine, it was an extremely bumpy place as planets and moons formed, smashed into one another, broke up, and re-formed. As the Solar System settled down, the rate at which impacts occurred dropped dramatically. But it did not drop to zero. Not only does our own Moon bear striking testimony to having been hit frequently over the ages, but so does every single body capable of retaining such evidence—planet, moon, or asteroid—that astronomers have been able to look at. So, too, does the Earth. Although weathering by wind and rain soon removes the most obvious features of impacts, some remain, such as the famous Meteor Crater in Arizona. And in living memory, a large area of the Siberian forest around Tunguska was destroyed by an asteroid impact, in 1908. All this knowledge has been in the hands of scientists for many years. The problem for these scientists has been when, how, and with what aim in mind to bring this information to wider public attention, or at least to the notice of policy makers who might act on it.

Every time the Earth expects a good cometary display, there is a rush of books, articles, and broadcasts about the facts and fantasies surrounding comets. They are harbingers of doom; one (now identified as Halley's comet) features prominently in the Bayeux tapestry depicting the defeat in 1066 of the Saxon King Harold by the Norman invader William.[41] The brightness of Comet Hale-Bopp in 1997 meant that nearly everyone in the northern hemisphere could see it, even in the light-polluted skies of the major cities. Not missing a trick, newspaper astrologers made great play of Hale-Bopp in their columns. For some of their readers, fame, fortune, and romance were to follow; for the one in 12 with a particular star sign, disaster would be the only possible outcome. But while it is easy enough to dismiss the predictions of popular tabloid astrologers, there remains strong evidence that, over the ages, people have looked at comets as heralds of change, if not always of disaster. And many believe that there are good historical reasons for humanity to be concerned about these visitors from the depths of the Solar System. Put simply, comets pose a threat that we ignore at our peril, they say.

The risk from comets, according to astonomers rather than astrologers, is that they may crash into us as they rush in from deep space toward the Sun. This threat is augmented by the presence of other pieces of debris—such as asteroids—left over from the formation of the planets and their moons. Some of these move in orbits very close to that of the Earth, and some are in Earth-crossing orbits; all could potentially be pushed onto a collision course with our planet. As a result, there is a considerable body of scientists who say that there is a risk that towns, cities, entire civilizations, or even the human race could be wiped out as a result of a comet or asteroid crashing into the Earth. It is easy enough to quantify the energy of such an impact. In space, things move fast, so a typical impact velocity would be about 20 kilometers per second. At that speed, a London bus would cause an explosion equivalent to several kilotons of TNT, not much less than the blasts that devastated Hiroshima and Nagasaki; and a solid rock approximately 30 meters across would cause a megaton explosion.

One spur to activity in impact-hazard assessment was the publication in 1980 of a paper in *Science* by physicist Luis Alvarez, his son the geologist Walter Alvarez, and their colleagues Frank Asaro and Helen Mitchell. On the basis of the analysis of a thin layer of clay that lies between the end of the Cretaceous and the beginning of the Tertiary geological period, this paper proposed that the age of the dinosaurs had been brought to an end as a result of the Earth colliding with a comet or asteroid some 10 kilometers in diameter.[42] Alvarez and co-workers claimed that this unusual clay layer, found worldwide, was the signature of a huge cloud of dust and debris, containing vaporized asteroid, which the impact explosion caused. This cloud, they said, would have blotted out the Sun, producing what became known, in 1980s-speak, as a "nuclear winter," killing plants and disrupting the food chain. Although the Alvarez hypothesis was by no means universally accepted (and it still has many scientific opponents), it was quickly followed by other studies that claimed to see a periodicity in the way in which large numbers of species simultaneously died out—events known as "mass extinctions"—and wondered if cosmic impacts were responsible for all of them.[43] Scientists began to investigate very seriously the threat to civilization from chance collisions with a stray comet or asteroid. In the early 1980s, NASA established a "Spacewatch" committee chaired by Eugene Shoemaker of the U.S. Geological Survey,[44] who had identified Meteor Crater as being the result of an impact rather than an extinct volcano. The public, too, were increasingly exposed to ideas about impacts as scientists involved in the multidisciplinary debate about the Alvarez hypothesis resorted to the general media to advance their own claims and refute those of others.[45]

During the early 1990s, as the Alvarez hypothesis gathered more evidence and gained greater acceptance, the U.S. Congress wanted answers to questions about present—rather than geologically ancient—dangers from space. In the defense industry and the Pentagon, there were those who argued that, as the Soviet Union crumbled, "rogue asteroids and doomsday comets" might provide the external foe necessary to justify the continuation of high-spending projects such as the "Star Wars" Strategic Defense Initiative. Nobler motives were also invoked: "only a threat from beyond the earth could unify the quarrelsome human species," commented veteran science fiction writer Arthur C. Clarke. "It may indeed be a stroke of luck that such a threat has been discovered, at just the period in history when we can devise the technologies to deal with it."[46] As a result, the American scientific community put around $1.6 million into a series of workshops and research papers looking at hazards to the Earth from impacts.[47]

At one of these workshops, British astronomer Victor Clube, a long-time proponent of the idea that past civilizations were brought to an end as a result of asteroidal or cometary impacts, remarked:

> By concerning itself and NASA with the hazards that space offers to Earth-bound civilization, the U.S. Congress has unwittingly revived the issue at the heart of theological debate these last two thousand

years, namely . . . whether the "revolutions of an invisible circulation in space sometimes affect the Earth . . . ." This is by no means a remote astronomical threat of the kind that is fashionable among scientific hedonists, such as the red-giant phase of the Sun, or a supernova explosion, or the universe collapsing in on itself to produce another supposed Big Bang; *rather, it is a hazard that is immediately at hand.* The probabilities imply an event tomorrow, next year, next century or the next millennium.[48]

At another workshop, Arizona astronomer Eugene H. Levy highlighted the responsibilities facing scientists investigating the cosmic hazards:

> The subject . . . is the last residual tail of planetary accretion: objects orbiting the Sun which occasionally collide with Earth. Only in the past few years has the reality and magnitude of this hazard been appreciated . . . . I believe that we scientists have a crucial role to play, a critical responsibility to fulfil. It is very important that the informed community of scientists participates in defining the realistic extent of the hazard and in defining sensible steps, if any, to take in response. It is also important for the scientific community to help weigh the hazards against the risks and dangers associated with conceivable protective measures.[49]

One of the scientists' proposals was the establishment of a Spaceguard Survey that would make a catalog of all potentially Earth-crossing (and therefore hazardous) bodies of more than a kilometer in diameter, at a cost of $10 million per year for 30 years[50]—though many doubted that it could reach this goal in so short a time. One problem facing the scientific community in trying to mobilize public and congressional support for such a scheme was how to demonstrate that there really was a threat to human life from astronomical impactors—after all, no-one in living memory had died from being struck by a piece of space debris. Another difficulty was how to cast risks averaged over astronomical timescales, of thousands, millions, and even billions of years, into the timescales of human politicians who tend to consider themselves far-sighted if they can see beyond the next election. According to astronomers Clark Chapman and David Morrison, there was a one in 10,000 chance that a large asteroid or comet would collide with the Earth during the next century.[51] As well as the immediate local destruction, such an event, they calculated, would bring climatic and social changes on so great a scale that food production would crash and one-quarter of the world's population would die in the famines and conflicts that followed. That meant that statistically "each typical person [in the United States] stands a similar chance of dying in an asteroid impact as in an aeroplane crash or in a flood." But, Chapman and Morrison admitted, public perceptions were very different:

> The nature of the impact hazard is unique in human experience. Nearly all other hazards we face in life actually happen to someone we know, or at least they are reported in the news. In contrast . . . our personal expectation of dying from an impact is extremely small, in spite of a surprisingly high level of statistical risk.

But in the summer of 1994—just a few months after the publication of Chapman and Morrison's article—an event occurred which put the image of a cosmic collision very firmly into the public consciousness.

Comet Shoemaker-Levy 9 had been discovered in March 1993 by Carolyn Shoemaker, David Levy, and Carolyn's husband Eugene Shoemaker, the chair of Spacewatch. From the outset it was an extraordinary body, resembling a squashed snowball. Soon it was realized that the comet was, in fact, composed of more than 20 individual fragments and that all of these would collide with the planet Jupiter during one week in the following July, causing explosions equivalent to hundreds of thousands of megatons of TNT. Given more than 15 months' notice, the astronomical community could gear up to a major campaign to observe the crash, using all the Earth-based and satellite-borne observatories available. Media interest was limited at first: in the British press, for example, just a handful of articles appeared during 1993, while the world of astronomy was making frenzied preparations. As the Impact Week approached, however, journalists—faced with politicians taking off for their summer holidays and looking for some respite from the soccer World Cup—began to take an interest in what the stargazers were up to. This, in turn, created a public space in which issues around cometary and asteroidal hazards could be discussed. Broadcasts and articles aimed at both children and adults linked the impending crash on Jupiter to the Alvarez theory of the death of the dinosaurs. Others went further and gave space to "extreme catastrophists," such as Victor Clube. Under the headline "Serial Killers from Heaven," Clube warned readers of the *Guardian*:

> The fact of collisions is recognized, but the message scientists are trying to get across to the public is that they don't matter. I think this is wrong, I think it is immoral, it's disgustingly wrong.
>
> Very roughly, these events are happening every 300 years. We have gone a couple of hundred years without anything happening, that's why we are so ruddy ignorant at the present time. That doesn't mean it is going to happen tomorrow, or in 10 years, or in 100 years,

but if you are reasonable you will educate people into recognizing that
there is a one in four lifetime chance for everybody alive today.[52]

To the annoyance of scientists on all sides of the "catastrophe" debate,
media coverage was also given to an assortment of quasi-religious cranks, the
most extreme of whom thought Shoemaker–Levy 9 would cause Jupiter itself to
explode and engulf the Earth, as God took His revenge on His wayward creation
Man. In May 1994, the "Polish astronomer Sofia (Religious name Sister Marie
Gabriel)" took out full-page advertisements in the British national press headed
"WORLD NEWS FLASH!" and addressed to British media organizations and to
world leaders such as the Pope, Presidents Clinton, Mitterand, and Yeltsin, and
Queen Elizabeth. Sofia claimed to have predicted the Jupiter impact long before
the astronomers and stated that it was a warning from God that the same fate
would befall the Earth if its inhabitants did not change their ways. We should reduce crime ("by copying Saudi Arabia's successful system of law and order"), destroy all pornography, ban all violence on television, stop killing old people
(according to Sofia, the health authorities do this, to save money), dress modestly,
end cruelty to animals, and stop all wars immediately. Sofia insisted: "People
must . . . beg God for mercy on their knees to stop the fireball asteroid."

The July 1994 impacts on Jupiter were spectacular. The Hubble Space Telescope relayed a series of pictures which showed rapidly expanding fireballs and
impact scars. Other telescopes produced equally dramatic images, which raced
their way around the world via the media and the Internet. The impact on the public psyche was enhanced by superimposing some of the larger impact scars on our
own Earth; a direct hit on the White House would have resulted in just one hour
in a cloud of dust and debris extending all the way to Antarctica. The cometary
impact followed by a "nuclear winter," as proposed by Alvarez and his colleagues
to account for the death of the dinosaurs, took on a chilling reality: the underlying
message was if Shoemaker–Levy 9 had happened here, we could be going the
same way. At that point, it looked as if those scientists who wanted governmental
support for a systematic assessment of the risk from cosmic impacts had, literally,
been given a heaven-sent opportunity to make progress. On July 21, toward the
end of Impact Week, *Nature* reported that the "greater than expected impact of the
comet Shoemaker–Levy 9 is set to intensify the debate over what steps, if any,
should be taken to deal with the prospects of such an object striking the Earth . . . .
Astronomers said this week that the Jupiter incident is likely to bring new impetus
to the Spaceguard proposal."[53] Clark Chapman, a supporter of Spaceguard, was
quoted: "It doesn't change the odds of a collision with Earth, but it makes the

prospect more real." And emphasizing the importance of the public's perception of the risk, Jet Propulsion Lab scientist Paul Weissman said: "It has demonstrated to people that comets do run into planets." The U.S. Congress set up a committee, again under the chairmanship of Eugene Shoemaker, to come up with a way of implementing the Spaceguard proposals. There were other scientific voices raised, however, expressing concern that the catastrophe camp might be overstating the dangers to Earth.

Evocative images are one thing, however, and a real appreciation of the statistical arguments is another. So quickly does the media pass from one story to another that, outside of scientific journals and the popular astronomy magazines, the release of the first definitive results from Shoemaker–Levy 9 in the spring of 1995 made hardly a ripple.[54] Attempts to keep alive popular interest in the comet, and the issues it raised, centered on a few television broadcasts and a number of books, which ranged from straight factual accounts to Australian comet watcher Duncan Steel's impassioned plea for politicians to stop "fiddling while Rome burns."[55] But, as far as Spaceguard was concerned, even these failed to keep the issues fresh enough in the public mind or high enough on the politicians' agenda for anything much to happen. A new, deeply conservative Congress lost interest in Eugene Shoemaker's committee; and Shoemaker himself was killed in a car crash in 1997. And, despite Steel's pleadings, the Australian government became less and less committed to even the limited Spacewatch program that he was involved with. The elaborate technology proposed for defending the Earth from comets was itself not without its risks. As astronomer David Morrison pointed out:

> Sooner or later, we may well have to assess the value and efficacy of the proposed defensive systems. It may also become essential to ask how the risks inherent in maintaining a nuclear asteroid defense system compare with the actual threat it is designed to mitigate.[56]

So it seems that hazards that drop on us from the heavens, with a regularity measurable only on astronomical timescales, which have only potential victims, no matter how many, and no guilty party whom we can sue (unless God has a bank account somewhere), are not seen as being something to worry about. The whirling heavens are part of the natural order of things, and if nature threatens to growl occasionally, be it in the form of a flood or a drought or a comet crash, we are inclined to react with stoicism, despite the high comparative risk. Perhaps the idea of an "act of God," while not necessarily in Sofia's terms, is still strong even in secular societies; acts of science, on the other hand, are not always met with the same equanimity. Statistically, the risk of a comet impact may be similar to those of earthly hazards that cause widespread concern; but the public perception is that it is not, and it is perceptions, not statistics, that make people vote. Politicians, with their hands on the purse strings and the levers of power, know this and act accordingly.

## LEARNING LESSONS

It is during episodes of risk that science and the public come together most urgently and acutely, and, if the cases outlined here are any indication, the interaction is to no-one's satisfaction. Are there, then, more general lessons to be learned from apples, beef, and comets?

The communication of risk is a challenge on many fronts. Specifying a risk precisely is a complicated task; the implications of a risk assessment can vary widely among different groups of people; the psychology of risk behavior is complex; and the social impact of risks is very difficult to define or predict. Scientists express risks as mathematical probabilities: most people find probability difficult to grasp, and the mass media tend not to deal particularly easily with numerical information. Newspaper stories tend to be self-contained and to stick to the point: so a risk story about nuclear power stations is unlikely to mention the risks from coal-fired stations that would enable readers to make some comparisons. But even comparing a risk to other risks that people already run does not always help. Risk scholar Baruch Fischoff and colleagues cite the following comparisons, expressed in terms of loss of life expectancy. Being unmarried costs men an average of 3500 days, while unmarried women lose only 1600 days. Smokers lose 2250 days, whereas people in dangerous jobs can expect to lose an average of 300 days.[57] But in life we cannot separate risks like this: how do the figures interact for, say, an unmarried soldier who smokes? Even statisticians have problems here. Murder takes 90 days off the average life expectancy; but it probably will not shorten *your* life by 90 days, because most people do not die of murder—and the ones who do, tend to lose rather more than 90 days. So this figure tells us something about the prevalence of violent crime, but how does a person relate this kind of information to their own individual life? Risk assessments apply on average, not to any particular case; and the mass media are little help: as media sociologists Sharon Dunwoody and Hans Peter Peters note in their study of risk communication, the mass media provide "information about society but not information about self."[58]

Comparing risks means mentioning at least two risks, which gives people twice as much to worry about. One of Bruce Ames's comparisons was that the risk

of cancer from the Alar in a glass of apple juice per day was 1/18 the risk of cancer from a peanut butter sandwich, and 1/1000 the risk of cancer from a glass of
beer[59]—very reassuring about apple juice, but rather discomforting about these
other American staples. Knowing about risks puts us at risk: if people know that
something is a potential hazard, it will have a real adverse effect on their lives,
even if the accident never happens. A study published in the *British Medical
Journal* in 1997 reported that people who participate in health screening programs suffer more illness after being screened—an effect that the researchers put
down to the stress of thinking about the possibility of disease.[60] Other studies
have indicated that people's attitudes to scientific risks become more negative
when there is more news about risk, irrespective of whether the news is good or
bad.[61] The study by Friedman and colleagues of Alar coverage demonstrated this
point: the coverage was not alarmist, but people were alarmed by the coverage
because there was so much of it.

The mass media are largely responsible for communicating risks to the public, particularly in times of crisis: in a world divided up into experts and laypeople, the mass media are often the only points of contact. Media theorists suggest
that the mass media do not tell people what to think, but rather what to think
about—they set the agenda. Risk is an ideal subject for the mass media because it
is something to think about even though nothing has actually happened: risk
makes news out of something that only might happen, and so it relieves journalists of the necessity of events. The mass media are interested in risk because danger is dramatic. Of course, the drama is in the risk event happening; that is, if
there is a chance of getting cancer from apples, it is "getting cancer" that makes
the headlines, not the more likely outcome of "not getting cancer." And the event
itself tends to be emphasized at the expense of its history or context, so that a
death from CJD might have been reported without any mention of the patient's
medical history, or any epidemiological information about the disease.

News needs "threshold"—big stories. A comet that might wipe out the
human race is news, even when it misses us by several million miles. Scientists
who warn that bits of space junk can be a danger to civilization may appear to
make progress in pressing their case for funding Spacewatch and Spaceguard programs. But once the immediate scare has died down, the numbers they wield concerning the probability of a direct hit—once in a million years, or once in 10
million years, depending on whom you talk to and when—seem just too large and
just too remote to maintain the pressure. Even if the Earth were to be visited by a
"cosmic catastrophe" every 100,000 years, politicians would say "that's a lot of
elections," which, for them, contain far more likelihood of a wipeout.

In dangers to humans, the biggest danger is death; so newspapers empha-size deadliness. In the news reports of an outbreak of botulism poisoning in Britain in 1989—it was traced to a hazelnut flavoring used in yogurt—there were many references to "killer bug" and "deadly yogurt," even though no one actu-ally died. But was this coverage irresponsible or alarmist? After all, the sensible precaution would seem to be to avoid hazelnut yogurt, which is a small enough sacrifice for most people—and not dying of botulism is surely adequate com-pensation. As Friedman et al. noted in the Alar case, sensational coverage is seen by many as a sign of poor journalism, and as compromising the message. But re-search suggests that consumers of the mass media recognize sensationalism when they see it: in a study of attitudes to information about food risk in Britain, respondents said that they trusted television documentaries and quality newspa-pers the most, then the government, the food industry, and environmental groups, and that they trusted tabloid newspapers least of all.[62] Risk scholars agree that the more sensationalist media are not the ideal forum for the resolution of con-flicting agendas. If the point is to solve the problem rather than to score points, then emphasizing extreme positions and personalizing conflicts is counterpro-ductive. As Rodgers points out, broadcasts about the dangers of Alar on prime-time television tend not to "facilitate a consensus-building dialogue on the underlying issues of policy."

Risk communication scholars recommend that any statement of risk should have two major features: it should give the likelihood of the event, and it should give its potential effect. For example, the chances of a power station exploding might be very small indeed, but the consequences could be extensive; a car crash is more likely than a train crash, but more people could be injured in the train than in the car. A theoretical danger, such as toxicity, should be separated from the risk that people will be exposed to actual danger: plutonium is extremely poisonous, but there is not much of it about. But when the public question a risk assessment, they are not simply questioning the size of the risk—they are trying to decide, as only they can, whether the risk is acceptable to them, and what to do about it. When the public make these decisions, they not only use knowledge, from a vari-ety of sources, but also make value judgments about acceptability in their per-sonal and social worlds. And as sociologist Friedhelm Neidhardt points out, scientific expressions of risk tend to concern particular effects such as a number of deaths or the incidence of a disease, whereas the public's assessments of risk tend to survey a rather broader spectrum of possible impacts and effects—what Hornig calls an "expanded vocabulary of risk"[63]—such as the kind of people who are most likely to be affected, the impact on employment and on the environment of any policy changes, or the possibilities for individual action to avoid or reduce the risk.[64] For example, Alar was alleged to present a particular risk to children; and CJD, normally a disease of the elderly, became a much greater cause for concern when it appeared to be affecting younger and poorer people.

Many factors involved in lay risk assessments are impossible to quantify or even to articulate in specific terms. For example, Jasanoff suggests that the public reaction to BSE was provoked not only by the risk to people's health, but also by the revelation that cows were not quietly munching grass in the meadow as nature intended but were instead being fed with the minced offal of sheep. As with the use of Alar to change the ripening patterns of apples, the situation was unnatural, and unseen hands were interfering in unexpected ways with the food we eat. Such knowledge can be unsettling, personally and socially; and no amount of risk information about CJD could have calmed this unease.

It is this broader and deeper approach to risks—a more richly informed assessment, socially and culturally, and one based on often unarticulated beliefs about how the world is or should be—that looks irrational to scientific eyes. But as Neidhardt points out, "with regard to social sensibility the public communication is obviously superior to the reductionist practices of science." Sociologist Ulrich Beck goes a step further:

> The non-acceptance of the scientific definition of risk is not something to be reproached as "irrationality" in the population; but quite to the contrary, it indicates that the cultural premises of acceptability contained in scientific and technical statements on risk *are wrong*. . . . the scientists . . . serve as judges of the "irrationality" of the population, whose ideas they ought to ascertain and make the foundation of their work.[65]

So while scientists may be able to calculate a risk, it is the public, Beck suggests, who know what is and is not an acceptable risk. The scientists' assessments are incomplete, for they are simply scientific, while the public's reactions embrace not only the scientific but also the broader, less readily defined qualities such as "environment" and "culture" that contribute in such large measure to the experience · of life and society. According to Neidhardt:

> In a highly differentiated society, where . . . medicine, politics [and] science . . . are specialized in order to serve citizens with high professionalization, the public is the place where the social acceptability of their output can be discussed, examined and evaluated. . . . The output of professional work has to match up to common sense before becoming implementable.[66]

The locus of common sense is also up for grabs in risk situations: in these polarized debates, each party claims its own position is common sense, while its opponents' is irrational. In the case of BSE, food minister David Maclean bracketed common sense with science, but Jasanoff reported a turn to common sense when science and authority failed. Perhaps people take up extreme positions in risk debates—they boycott apples or declare "beef is absolutely safe" and condemn anyone who thinks otherwise—because such positions allow a clear and unambiguous frame of mind. In times of uncertainty, it is a clear message that wins the day, both for personal psychological comfort and for social action: it allows us to know both how to feel and what to do. Risk messages are difficult messages, and not just for laypeople; and faced with this difficult problem, the tendency is to turn an intellectual problem into a practical and moral problem: what is to be done, and whose responsibility is it? According to Neidhardt, "in the end, people want to know who has to do what to whom."

Hazards most often occur where the technology is new: they exist because scientific knowledge is provisional and inevitably incomplete. In this situation, even the experts are dealing with a barrage of unknowns and a range of often conflicting agendas. Yet it is the public whose ignorance is more often highlighted. As sociologist Brian Wynne explains:

> . . . legitimate public ambivalence and resistance to expert presumptions about the framing of risk issues was first interpreted as simple ignorance, then "misunderstanding," and latterly as a naive wish for an impossible "zero-risk" environment. The constructions of the public ignore sociological evidence that shows that people are by no means naive about the existence or complete eradication of risks, and points rather to scientists' unacknowledged insecurity about recognizing the conditionality of their own knowledge . . . .[67]

Media scholars JoAnn Valenti and Lee Wilkins also call for more openness from scientists:

> The scientist engaged in professional activity has specialized knowledge—and an understanding of that knowledge that . . . it is his/her responsibility to communicate. Further, that responsibility for communication is part of the implied promise that scientists make not only to their employers but also to society at large which has given them a position of special responsibility.[68]

Valenti and Wilkins argue that both scientists and journalists have an ethical responsibility to disclose not just the risk assessment itself, but also as much information as possible about how the assessment was made; they should also include the social and qualitative elements of risk assessment that are important to the public. Since both lay and expert rationalities are at work when people interpret risk messages and act upon them, Valenti and Wilkins suggest that both lay and expert perspectives should be brought to bear on the construction of the risk information. Risk communication, they suggest, should be a dialogue between all parties concerned:

> Public acceptance alone cannot legitimate public policy decisions that are not supported by the scientific community. But the intellectual history of the field of risk communication indicates that it is the scientific community which has often been considered the arbiter of truth. We assert that the public has an equal stake in the truth-framing and policy-making processes.

To achieve, then, a risk communication dialogue that functions in the relevant cultural environment and enables non-experts to make choices about risks, Valenti and Wilkins have set out what they describe as an "ethical risk communication protocol for science and the mass media."[69] The protocol makes demands on scientists, journalists, and public relations practitioners.

Scientists dealing with risk have a responsibility to communicate more than just the bare facts of the case—more than simply a probability statement, or a straightforward reassurance. Where the data for the assessment came from is relevant; as are the causes, effects, and implications of the risk, and whose interests are at stake. As Friedman and colleagues observed in their study of the Alar scare, it is part of the journalistic ethos to report not just the facts but also the truth about the facts; and in the case of the Alar reporting, it was this "truth about the facts" that was missing. Of course, including all this background information would have taken up the kind of space that newspaper articles just do not have. The argument of Friedman and colleagues, though, is that because risk is complex, so ought to be the coverage: they suggest that features should be attached to news articles or that news articles should guide concerned persons to web pages. So journalists should not simply inform the public but should give their audience the chance to inform themselves. Avenues for further investigation should be opened up; opportunities to obtain answers to questions should be made visible—be they telephone information lines, further reading, or addresses of advocacy groups. Sofia's asteroid warning advertisement actually fulfills this criterion at least: she offers the telephone number of the Scientific Forecasts Society, a charity which aims "to explain to the public the divine meaning behind major cosmic events through the press and television."

The public can be a source as well as a recipient of information for journalists. The layperson is expert in one key area: that of the personal experience of living in a risk environment—the public knows what it wants to know. In her study of BSE, Jasanoff comments on the value of a more democratic approach to risk information:

> . . . lay questioning, however ignorant or ill-founded, might have led to deeper reflection on the limits of expert knowledge and, in turn, to more collaboration among citizens, scientists and government about how to manage the multiple uncertainties of mad cow disease.

Public relations practitioners should also be aware of their broader responsibilities, suggest Valenti and Wilkins: somewhere within their possibly conflicting responsibilities to their client, the media, and the public, they should make clear not only what the message is, but also whose message it is. According to political scientist Philip J. Frankenfeld, the important question for citizens in a technological society is not "how safe is it?" but "who controls what, and by what right?"[70] Valenti and Wilkins also suggest that public relations agents should provide the public with the means for engaging themselves in a continuing dialogue: the aim should be "to persuade the targeted public to consider the provided information, not to compel people to believe and then act on a particular 'truth'."

These strategies offer ways of reducing the information problem. But do they solve the risk problem? According to Wynne, information may be a key factor in the dynamics of risk situations such as the scares over Alar or BSE, but it is often far from sufficient for the resolution of these episodes in the relationship between science and the public. Wynne argues that the crucial dynamic in risk communication is trust: reactions to risk, he suggests, owe more to the public's experience and perceptions of the institutions involved than they do to any understanding of a technology and its risks.[71] Hornig's research corroborates this: it indicates that people's level of concern about risk is determined not by what they know about the risk, but by how they feel about the agencies controlling the risky technology.[72] Two readers' letters published during the BSE scare endorse this point:

> . . . the whole point about the recent evidence is that we (the general public) just don't know. We don't know how much this Government (not renowned for its openness in these matters) knows either, nor can we rely on it to alert us quickly to any danger, no matter how remote. Assertions from . . . other parties with vested interests as to the safety of British beef are hardly unbiased.[73]

> As a member of [the] "populace at large," I do not believe that the gov-
> ernment has my best interests at heart. . . . This lack of trust, unfortu-
> nately, applies also to scientists . . . . I listen with increasing
> apprehension to scientists publicly denigrating each other's theories
> and results. . . . As members of this unfortunate mass we have to trust
> our pessimism, for to whom else can we turn?[74]

In a science as difficult and emotive as a risk assessment, trust is all the more important: according to Neidhardt, "when people can not understand they have to believe. And whether they believe or not is a matter of trust."[75] Trust may even make information redundant: workers who trust their bosses, or patients who trust their doctors, may choose not to know about risks and to place their destiny in the employers' or doctors' trustworthy hands.[76] But this trust is not easily or readily won: for while information can be mobilized instantaneously in a crisis, trust is much more difficult to generate at a moment's notice.

In the case of BSE, Jasanoff points out the peculiarly personal repositories of trust in British society, which she contrasts with the institutionalized trust of the United States. In Britain:

> . . . trust is created through embodiment in trustworthy people . . . .
> Over and above any demonstrations of technical competence, such in-
> dividuals have proved their right to represent the public interest
> through years of devoted service. . . . People who have attained this
> status can be said, with little exaggeration, to constitute an elite tier of
> civic virtue that stands, and is seen to stand, above self-interest and
> even party politics.[77]

When the British trust their advisers, then, there is no need for the agencies these advisers constitute to give an account of how they arrived at their recommendations: the kind of information that might enable the public to evaluate risks for themselves is superfluous, because the trustworthy agency has done this for them. When John Gummer asserted that "My wife eats beef; my children eat beef; and I eat beef: that is everyone's absolute protection," he was giving the British public exactly the sort of reassurance that they would have required had they trusted him. But when the trust is gone, this sort of statement becomes meaningless and leaves the public with nothing from which to determine their own course of action. In the United States, on the other hand:

> Persons do not command much faith . . . . Instead, trust is reposed in
> formal processes, such as rule-making and litigation, and in styles of
> reasoning that ensure the transparency and objectivity, if not the wis-
> dom, of governmental decisions. . . . US decision-makers have sought

more and more to justify their actions through seemingly objective, mathematical assessments of risks, costs, benefits and, lately, even environmental justice or equity. Paradoxically, policy-making in this most open and transparent of political cultures has come to depend most aggressively on abstruse, technical and distancing expert discourses.[78]

Britain, then, seems to suffer from too much reliance on the presumed trustworthiness of people in authority, and from too little provision of information for the public to use to make their own decisions when the trust evaporates. The United States, on the other hand, relies on an unassailable and impersonal machinery that produces information which it claims is trustworthy on the grounds that the machinery is reliable; it is, however, a machine that can produce conflicting information, and one whose workings are opaque to most laypeople. In neither country do the pronouncements of the experts take account of the broad spectrum of criteria against which the public judges the acceptability of risks. As Hornig emphasizes:

> No amount of information on the probabilities of harm—however phrased—will serve to create a favorable climate of public opinion unless social context issues are also addressed. Where they have not been addressed (either in actuality or in media representations of that actuality), public pressure to address them is likely to continue, and lay evaluations of risk will continue to be responsive to the deficit.[79]

What emerges from these arguments, then, is a call for the social negotiation of risk: a dialogue which constructs risk communications that serve the needs and interests of all parties in every relevant part of their lives. Risk communication should not be "one-way propaganda to manipulate acceptance of someone's assessment of acceptable or unacceptable risk,"[80] but a social process based in mutual trust that enables a community to use information to determine its own risks and its own engagement with risky technologies.

**Chapter 8**

# Science in Museums

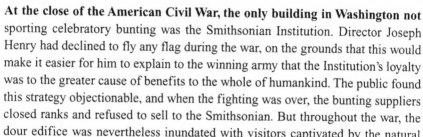

**At the close of the American Civil War, the only building in Washington not** sporting celebratory bunting was the Smithsonian Institution. Director Joseph Henry had declined to fly any flag during the war, on the grounds that this would make it easier for him to explain to the winning army that the Institution's loyalty was to the greater cause of benefits to the whole of humankind. The public found this strategy objectionable, and when the fighting was over, the bunting suppliers closed ranks and refused to sell to the Smithsonian. But throughout the war, the dour edifice was nevertheless inundated with visitors captivated by the natural history collections that Henry had for so long tried to keep out of the Smithsonian: a physics experimentalist and teacher, he had accepted only reluctantly the often flea-ridden and poorly preserved collections pressed on him from all sides. The public, however, wanted to see birds and rocks and flowers; and the soldiers garrisoned in Washington found themselves, like so many people at a loose end in a strange city, in the Exhibition Hall pondering the wonders of nature.[1]

The problems of the Smithsonian in wartime were an early taste of the challenges facing many other museums 150 years later. In times of fluctuating political cultures and rapid social change, facing tensions between the aims of scholarship and the goal of public communication, and increasingly committed to serving visitors it understands only poorly, today's museum is, according to media studies professor Roger Silverstone, "no longer certain of its role, no longer sure of its identity, no longer isolated from political and economic pressures."[2] Yet museums have been key players in public-understanding-of-science "movements," and they are learning to negotiate loyalties to and dependencies on governments, the scientific community, and the public. In the next chapter, we look at how the wider scientific community is responding to the challenge of communicating with the public; this chapter looks at the rise and evolution of science museums and at their role as the most public of the institutions of science.

## THE ORIGINS OF SCIENCE MUSEUMS

The museum is a 17th-century innovation in the Western world, born out of the cabinets and collections of merchants and explorers. Among these were John Tradescant's collections, which were bequeathed to his friend Elias Ashmole. Ashmole displayed the collections at "Tradescant's Ark" in his home in Lambeth, London, until February 1683; then the 26 crates of natural curiosities traveled by barge and cart to Oxford, where the Ashmolean Museum—the first public museum in England—opened its doors a few months later.

Thus, the Ashmolean began life as a science museum, with a broad Renaissance perspective of science as culture: its aim was to further " 'the knowledge of Nature' acquired through 'the inspection of Particulars'." It consisted of a laboratory, exhibited collections, and a lecture hall, and all the sciences were taught there—indeed, the lectures at the Ashmolean were the origin of science in the University of Oxford.[3] The museum became a museum of the history of science in 1924, in order to preserve the past as science hurtled ever more rapidly into the future, scattering ephemera in its wake.

In the 18th century, the philosophy of the Enlightenment could be put into practice in public museums. The French zoologist Georges Cuvier recommended museums to Napoleon as vital to the future health of the sciences and of the nation:

> Your Majesty will no doubt order the maintenance and growth of the gardens, cabinets and other collections in the regions. These teaching materials speak ceaselessly to the eye, and inspire a taste for science in young people . . . the present era will be reproached if we do not conserve for the future these sources of so many advantages.[4]

Cuvier was for many years the director of the Jardin des Plantes, a botanical garden in Paris that had begun life as the French kings' medicinal herb garden. The Jardin des Plantes flourished as a public institution in the 18th century and remains committed to public enlightenment, as it was in Cuvier's day; it comprises not only the gardens but also museums, lecture halls, and a research institute. The naturalist Louis Agassiz used the Jardin des Plantes as the model for the Museum of Comparative Zoology at Harvard, which he founded using grants and bequests gathered during the 1850s. Agassiz was a supporter not just of Cuvier's institution; he also supported Cuvier's increasingly unpopular ideas about evolution and speciation. Historian Robert V. Bruce has described Agassiz's museum, which opened in 1860, as "a temple to Cuvier, an anti-Darwinian fortress"; and while its allegiance may have shifted, the museum, as the long-standing academic home of the evolutionary theorist and writer Stephen Jay Gould, remains a center for the public debate of Darwinism. Agassiz's museum was privately funded, but debates about public money were not far off: when the Smithsonian opened its Zoological Park at the end of the 19th century, a Tennessee congressman denounced the waste of taxpayers' money on building a home for "all the creeping and slimy things of the earth," when many Americans were themselves without homes.[5]

Personal missions and enlightenment philosophy still play a part in the development and survival of museums. But a new force entered the arena in the 19th century: the "great exhibitions" that celebrated a nation's power and technological prowess. The British Empire's "Great Exhibition" of 1851 also gave space to other nations—a prospect that the British Establishment had found most unsettling. The Exhibition was the suggestion of Albert, the Prince Consort, and as a Saxon he was himself unpardonably foreign in many English eyes. His ambition to invite foreigners onto English turf had generated hate mail and was denounced in the House of Commons: England would be overrun by spies and revolutionaries, who would subvert and corrupt the English, steal their inventions, and divert them from their love of their Sovereign and their God. But Albert persevered, believing that some insight into other countries' industries could only stimulate the British. His Exhibition eventually covered 21 acres in South Kensington, London, and 14,000 exhibitors showed their wares to 6 million people.

The Great Exhibition of 1851 showed not only that the British Empire was far advanced over rival nations in industry, but also that it was backward in science education and training.[6] One consequence of this realization was the establishment of a government department for "Science and Art." This department organized the objects left over in 1851 into collections for the new South Kensington Museum, which was funded by the £150,000 profit from the Great Exhibition. The new museum opened in 1857; it split into the Victoria and Albert Museum, for decorative arts, and the Science Museum in 1928. The Science Museum was intended to record achievements in science and, by displaying these in the form of technological objects, to inspire further efforts: it should "serve to increase the means of Industrial Education and extend the influence of Science and Art upon Productive Industry."[7]

International exhibitions had their most spectacular incarnations in the United States during the second half of the 19th century, when grand, ambitious "world's fairs" rose up in different parts of the country—Philadelphia in 1876 attracted 10 million people, and 21 million people went to Chicago in 1893. The fairs were part amusement park, part trade show, and part educational outreach. President William McKinley, who was assassinated at the World's Fair in Buffalo in 1901, spoke of the power of the fairs:

> Expositions are timekeepers of progress. They record the world's advancement. They stimulate the energy, enterprise, and intellect of the people and quicken human genius. They go into the home. They broaden and brighten the daily life of the people. They open mighty storehouses of information to the student. Every exposition, great or small, has helped this onward step.[8]

The world's fairs were also, according to historian Robert W. Rydell, "triumphs of hegemony," reinforcing and guiding the social order on many more matters than just the public image of science and technology.[9] The huge numbers of people who attended—100 million between 1876 and 1916—found there a single vision for their increasingly plural and rapidly changing society. It was not a vision that was attractive to everyone; yet the fairs' claims for America's assets and prospects, juxtaposed with their depictions of other cultures, offered visitors both hope for the future and a sense of what it meant to be an American.

The Chicago fair of 1893 featured a Ferris wheel 264 feet high that seated 2000 people. Though it could not outdo for sheer size the Eiffel Tower, built for the Paris exposition in 1889, the wheel's claim to technological fame was that it ran on the largest one-piece axle ever forged. Huge pavilions connected by waterways could be reached by electrical boat, and the international displays were described in one contemporary account as "a cacophonous confusion of advanced cultures, converted heathens, peculiar tongues, and queer importations."[10] At a time when electricity was still a novelty available only to the rich, the people who flocked to Chicago could visit the three-and-a-half acre Electricity Pavilion and see the Tower of Light lit up by 18,000 light bulbs, a high-frequency coil that produced sparks five feet long, a telephone that transmitted the voice using light beams, and artificial lightning. Electricity for the fair was generated on site by Westinghouse, and more was required than usually ran the whole city of Chicago.[11]

When the Chicago fair closed, its Palace of Fine Arts was turned into a natural history museum, but after a few years it fell into disuse. The Palace was saved from dereliction in 1929 when it was restored as a museum of industry. Charles

Rosenwald, chairman of the Sears, Roebuck Company, had been inspired by a visit to the Deutsches Museum in Germany, which had captivated his 8-year-old son. Rosenwald mustered the money from the business community in Chicago, and the new Chicago Museum of Science and Industry opened in 1933. It benefited from the world's fair that was held in Chicago again that year, and which conveniently left behind a selection of exhibits.

By the end of the 19th century, then, the pattern for the great science museums was established, and many new institutions were to follow. These were grand institutions: expressions of the pride of nations in their technological achievement, and statements of the natural order, both scientific and social, by both scientists and governments. They displayed the material culture of science—tractors, oscilloscopes, beetles, rocks—to a public that would not necessarily understand, but that would be impressed and inspired nevertheless. In the United States, the museums tended to represent the present and to have close links to the curriculum; they consigned the past to museums of history. In Britain, the museums tended to record past achievements, and to ignore the ever-changing present. Exhibits were to be looked at, though staff might occasionally demonstrate the machinery, and there might be a button to press to set an object in motion. While the museum community might reflect on changes of emphasis and style in the century since the Great Exhibition of 1851, for the public the museum experience was unchanging. But science museums were set for some surprises in the 1960s, the repercussions of which still challenge museum professionals today.

## THE RISE OF THE SCIENCE CENTER

Francis Bacon's *New Atlantis* was published in London in 1627, a year after his death. The pioneer empiricist was weary of the Ancients with their myths and mere theorizing, and wanted to get his hands on the real world. In the imaginary world of New Atlantis is the House of Salomon, wherein

> we make demonstrations of all lights and radiations, and of all colours; and out of things uncoloured and transparent, we can represent unto you all several colours; not in rainbows, as it is in gems and prisms, but of themselves single. . . . We procure means for seeing objects afar off; as in the heaven and remoter places . . . . We also have helps for sight, far above spectacles and glasses in use . . . . We also have sound houses, where we practise and demonstrate all sounds and their generation. We have harmonies which you have not, of quarter-sounds and lesser slides of sounds. . . . We have the means to convey sounds in

trunks and pipes, in strange lines and distances. . . . We imitate also flights of birds; . . . we have ships and boats for going under water, and brooking of seas; also swimming girdles and supporters. We have divers curious clocks, and other like motions of return, and some perpetual motions. . . . We have also houses of deceits of the senses; where we represent all manner of feats of juggling, false apparations, impostures, and illusions; and their fallacies. And surely you will easily believe that we have so many things truly natural which induce admiration, could in a world of particulars deceive the senses, if we could disguise those things and labour to make them seem more miraculous.[12]

The most immediate influence of *New Atlantis* was in the setting up of the Royal Society as an institution of experimentation and discussion. But Bacon's 17th-century house of the tricks, objects, and phenomena of science is often cited as the manifesto for the interactive science centers that, three centuries later, have transformed the experiences of museum visitors around the world over the last few decades.

As curator and historian of science Willem Hackmann notes, interactive science is not new: in the 18th century, "hands-on" demonstrations were an important component of a traveling lecturer's stock in trade; they were also a key part of the earliest university science courses.[13] But at the end of the 19th century, in England at least, objections were raised against experiments in science education, on the grounds that children should accept the word of scientific authority, rather than interact directly with nature themselves.[14] A change in attitude became apparent in the aftermath of Sputnik, when science museums, and the new science centers, looked to working models and interactive exhibits to engage the interest particularly of children and to stimulate some understanding of scientific principles. According to Sheila Grinell:

> In the late 1960s, . . . several institutions opened that elaborated on the concept of interactivity. . . . [and] eschewed historical and industrial collections in favour of apparatus and programmes designed to communicate basic science in terms readily accessible to visitors. These institutions postulated that displays and programmes carefully designed to provide first-hand experience with phenomena could captivate ordinary people and, in the best of circumstances, stimulate original thinking about science.[15]

Grinell worked at one of the pioneering institutions in this rediscovery of experiments: the Exploratorium in San Francisco, which opened in 1969. Its founder and guiding light, Frank Oppenheimer, had been inspired in part by the Children's

Gallery in the London Science Museum, which had opened in 1931, with buttons to push and handles to turn to activate machinery and demonstrations. Oppenheimer had worked alongside his brother Robert on the Manhattan Project and had left physics for teaching after being blacklisted in the 1950s for his political associations before the war. As a teacher, Oppenheimer developed classroom demonstrations to give his students some direct experience of physics; and he became convinced that the route to understanding was via the senses—not just sight and hearing, but also touch and smell, balance, and motion. Oppenheimer's vision of a museum that used the power of perception to access the natural world was realized in the Exploratorium, which was housed in a building that had originally been constructed for the World's Fair in 1915. In the Exploratorium, experiments and demonstrations, texts, and human "explainers," along with art, music, and dance, combine to engage visitors in the science. Oppenheimer's achievement became the model for science centers around the world: his staff even published what they called "cookbooks" that contained recipes for making exhibits and suggestions about how to explain them.

The San Francisco Exploratorium makes full use of what is now known as the "informal learning environment," in contrast to the ordered, didactic galleries of traditional science museums. The exhibits have a deliberate "string and sealing-wax" feel to them; the inclusion of the center's workshop within the main hall of the Exploratorium means that visitors may watch new exhibits being created and old ones being repaired. But that is by no means to say that the center is either ramshackle or unstructured. Visitors make their way through themed areas, on weather, for example, or sight and sound, experimenting as they go.

In Britain, psychologist Richard Gregory, who had worked with Oppenheimer to develop some exhibits about perception, started his own Exploratory project in 1979. The Exploratory moved into its present home in a disused railway station in the center of Bristol in 1988. As in the Exploratorium, the Exploratory's exhibits (which Gregory calls "plores," from "explore") are often rough and ready and are made to be touched. Initially, perception was the main area for the visitors to explore, but the Exploratory has expanded its interests to encompass a wide range of principles and science and technology.

These pioneering interactive science centers have spawned many imitations: indeed, one indication of the success of the Exploratorium's "cookbooks" is that the same exhibits can be found in science centers around the world. Free of the demands of amassing and conserving historical artifacts, science centers can be built anywhere, from scratch. Thirty years after Oppenheimer's experiment in interaction, there can be few nations that do not boast a science center, and in some countries a science center is de rigueur in even the smaller towns: in 1997, there were about 40 science centers in Britain and 300 in the United States.[16] Unlike museums, with their national and scientific allegiances, science centers tend to be rooted in the local community and to reflect local culture and interests. For example, the Fondazione IDIS in Naples, Italy, plans to build an employment agency alongside its science center, to encourage local people with an interest in science to consider scientific jobs. Countries with dispersed populations or limited resources find mobile solutions: traveling "science circuses" tour Australia and Spain. Science center professionals speak of the "science center movement" and have formed two very active professional associations: the European Collaborative for Science, Industry and Technology Exhibitions (ECSITE), which is based in Brussels, Belgium, and the Association of Science-Technology Centers (ASTC), which hails from Washington, D.C.

## SCIENCE MUSEUMS TODAY

Many traditional science museums have embraced the innovations developed by science centers and now offer interactive exhibits, either as part of exhibitions or as an exhibition in themselves. Museums and science centers also use many other techniques for communicating science to their visitors and for reaching their local communities.

The Franklin Institute of the State of Pennsylvania for the Promotion of the Mechanic Arts—now the Franklin Institute Science Museum in Philadelphia— was founded to demonstrate the interdependence of science and craft skill in the "mechanic arts." The Institute began running public lectures in the 1820s and has been touring neighboring schools to give demonstrations and lectures since the 1930s. In a tradition dating back to the 19th century in the United States and Britain, the Franklin Institute also offers "Museum to Go" science activity kits for teachers to take back to their schools. The Boston Museum of Science's outreach programs include Science-by-Mail, in which children are sent a science challenge and put in touch with a scientist "pen pal" who can encourage and advise. The New York Hall of Science rents out STARLABS: portable planetariums that teachers, who are also offered training by the museum, can set up in schools to demonstrate the basic principles of astronomy. Science North in Ontario, another pioneer of interactive science, was the first museum to actively encourage visitors to fall asleep: children armed with sleeping bags can now take part in science activities and then camp overnight among the exhibits in many museums.

The portrayal of scientists and users of technology by actors working with visitors in the galleries was begun by the Science Museum of Minnesota in the early 1970s and is now a feature of many museums around the world. As an interpretive technique, it has the unfortunate capacity to repel as many visitors as it attracts; but there are many who enjoy gallery drama. Graham Farmelo of the London Science Museum argues that gallery drama also introduces what is so often lacking in science museums: an exhibit that is also a person, who can show the human side of science with such simple tactics as discussing, for example, how uncomfortable the earliest bicycles were to ride, or how noisy the cotton mills were, or how difficult it was to sit still long enough to be photographed by a 19th-century camera.[17]

Collecting for science museums today is not a straightforward process. Distinguishing the ephemera from the momentous is increasingly difficult when the pace of change is so rapid, when the sheer quantity of scientific objects is overwhelming, and when mass production makes technology both enormously varied and, at the same time, commonplace. Increasingly, technology is black-boxed for consumption, and its outward appearance is neither explanatory nor attractive. Science has wandered off the everyday scale and concerned itself with the cosmic and the microscopic—neither of them easy to display. As museums scholar Stella Butler points out, collecting even the everyday material culture of 20th-century science can involve finding space for some very large objects.[18] The London Science Museum has fitted the landing gear of a 747 airliner into its entrance hall, but its collection of airplanes lives mostly in a field in the Wiltshire countryside. With bigger objects still—bridges or particle accelerators—the problem has been solved by turning the object itself into an exhibit: instead of being a small object in a large museum, many of the big institutions of 20th-century science and technology are large objects with a small museum attached. One example is the Jodrell Bank Visitor Centre, a small museum devoted to exhibiting one very large radio telescope. The Centre, at the University of Manchester, began life in the summer of 1964, when Jodrell Bank was frequently in the news in Britain for its observations and tracking of the space race. The university pitched a large tent in an effort to contain the chaos as tourists, passers-by, and astronomy enthusiasts—35,000 of them that first summer—wandered up to the telescope and wanted to know more about it. Thirty years later, the Centre boasts an exhibition, a planetarium, an arboretum, and a handful of staff and is financed by its 150,000 paying visitors each year.[19] Other universities have seen the importance of getting into the museum and science center business as an addition to their traditional academic activities. For example, the University of Leicester, which has a large space physics group, is the location for Britain's first National Space Science Centre, benefiting from the country's Millenium Fund.

Traditional museums were built on the contemporary technology of the 19th century: iron, steam, and coal. These industries are themselves now merely history in much of the Western world: rapid industrial change and economic turns of fortune in the postwar era have closed the coal mines and the steel mills. Some of these centers of science and community have become "heritage" museums, where technology and ways of life are preserved—creating jobs and leisure opportunities according to some, and bogging societies down in the failed technologies of the past, according to others.[20] But redevelopment has also brought opportunities for museums that started from scratch, unburdened by the inertia of traditional institutions. A redundant railway station in northern England became the Manchester Museum of Science and Industry, and a disused stretch of land on Baltimore harbor is now home to the Maryland Science Center. Nearby, the Center of Marine Biotechnology has taken over Pier 5. Established in 1985, it is a center for research and education, formal and informal; visitors can carry out their own experiments on remedying water pollution under the guidance of research staff.

Some new industries, particularly where there is public ambivalence about their activities, have taken to exhibiting themselves, rather than having their role interpreted by an independent museum. In Britain, the Sellafield Visitors' Centre is a museum-cum-science center attached to the nuclear power station formerly known as Windscale, in Cumbria on England's northwest coast. Sellafield has been a source of local and national concern ever since a leak of radioactive material occurred during a fire there in 1957. Plans to expand nuclear reprocessing and arrangements for storing radioactive waste regularly come under fire from groups ranging from local environmentalists to the Irish Government. During the school holidays, the Sellafield Visitors' Centre is advertised on national television with the slogan "where science never sleeps." Public relations manager Duncan Jackson believes that giving people the personal experience of visiting a place like Sellafield, and of sampling information from the range of media offered there, is a very powerful way of influencing public opinion: it provides opportunities for education, entertainment, and marketing.[21] The nuclear fuel reprocessing company BNFL, which runs Sellafield, states that "we have a duty to explain clearly what it is we do, to enable the public to make clear judgements about our activities rather than basing their views on misconceptions."[22] Tours of the power station itself and talks from current and former staff allow a closer engagement with the practitioners and substance of contemporary science and technology than most museums can offer. The Sellafield Visitors' Centre, which opened in 1988, has certainly been popular: it attracts 200,000 visitors each year.

## THE MUSEUM AS MEDIUM

The first museum devoted to technology was opened in Paris in 1799. Set up by decree of the postrevolutionary government, the Conservatoire des Arts et Métiers aimed to be a public storehouse of machines, books, and drawings, where people could learn about mechanics and industrial processes. The museum was housed in the medieval priory of St-Martin-des-Champs; and there it remains, two centuries later, in the rue St-Martin, where its soaring arches and vaulted ceilings provide the ideal suspension for a Foucault pendulum.

The Conservatoire is literally a church; but the metaphor of the museum as temple or cathedral to science is often used by both advocates and critics of traditional object-based institutions. In their accounts, key artifacts—Pascal's calculating machine, or Stephenson's Rocket—are enshrined in space and light or lovingly conserved against the odds as though they were holy relics or sacred icons. The vast spaces of grand exhibition halls shrink the visitor. The glass cases and restraining barriers proclaim the visitor unfit to approach the precious objects of science. Science in this vision of museums is awesome and marvelous. It is unassailable, powerful, and to be revered. It is science off the human scale and outside society. It is beyond understanding, but it is brought to earth in museums so that we may merely glimpse its triumphs. According to museum developer Melanie Quin:

> The construction of a museum is a declaration that the government and others want to influence public attitudes to science and technology, and to increase the standing of these subjects (and of scientists, engineers, etc.). The museum is a prestigious monument to that aim.[23]

Some people wander around science museums just as they wander around churches on the tourist trail, impressed simply by the architecture or the decoration. For some, a visit is an expression of faith; but for others, the feeling generated by a science museum is the unease of the unbeliever in a place of worship. Museums, rooted in the philosophy of the Enlightenment but born out of a 19th-century commitment to educate, stand as authorities in the transmission of the pure facts of nature. As Butler notes, science museums built at this time were to promote a particular status for science. Their development was part of the process through which scientists sought recognition as a profession, producing useful and authoritative knowledge. It is not surprising, then, that science is presented by museums as certain knowledge, arrived at through painstaking, objective research.[24]

And woe betide anyone who did not follow the party line. In Chapter 5 we mentioned the scientists' boycott of the publishing company Macmillan when it published Immanuel Velikovsky's controversial cosmology. In museums too, this episode had its impact. For example, in 1950 curator Gordon Atwater of the Hayden Planetarium in New York was forced to resign for mounting an exhibition that failed to damn Velikovsky with sufficient vigor.[25]

But as is the case for all institutions of science in the modern era, the authority of museums is open to challenge. The grand failures and unresolved disputes of science in the 20th century jostle for exhibition space with the triumphs of yesteryear, which rapidly look clumsy if not ridiculous. The atom and the double helix may still be icons, but they are also contentious objects, and the context of their display is debated. As cultures generally, and even the particularly monolithic scientific culture, move more toward a pluralistic approach to their membership and their ideas, so museums become less like cathedrals to mainstream culture—not least because mainstream culture becomes increasingly difficult to define. Instead of being places that define culture, museums become another medium to throw their interpretations of cultures into the public arena. As Joel Bloom of the Franklin Institute Science Museum has pointed out, "there is no such thing as a neutral museum exhibit."[26]

What emerges at the end of the 20th century, then, is the realization that not only is science constructed by human authors, but so too is the museum. The more adventurous museums are beginning to see themselves not just as carriers of the prescribed messages of science, but as authors of a text that is open to interpretation. The museum is beginning to recognize itself as a medium, and museum practitioners and analysts are seeing the museum experience as involving the same influences of authorship and framing and multiple interpretations and impacts as are found in the reading of a magazine article or the watching of a television program.

But unlike a magazine, which next week will be thrown away, or a television program, which may never be broadcast again, the museum is here to stay: an exhibition may stand for 20 years. This unchanging representation of science in the museum may not only misrepresent the changing face of science, but may also, according to anthropologist Sharon Macdonald, exaggerate the authority of the institution and its contents.[27] The longevity and fixity of exhibitions make controversy a particularly difficult topic for exhibitions: multiple points of view are difficult to display to even the most attentive and persistent visitor; and the essential dynamism of controversies is lost once the content of the exhibition is fixed. In disputed science, how is the visitor to know whose point of view is on display? Exhibition texts and displays are rarely ascribed to a particular author in the way that newspaper articles or TV programs are credited: exhibitions have the authority not of some individual (fallible) journalist whose reputation one can judge, but of the museum within whose invariably grand portals one is standing. Yet for all the social and intellectual authority of museums, their staff are not mere transmitters but active authors of the scientific culture they present. In museums as in other media, Macdonald emphasizes:

> . . . science communication involves selection and definition, not just of which "facts" are presented to the public, but of what is to count as science and of what kind of entity or enterprise science is to be. That is, science communicators act as authors of science for the public. They may also, however, by dint of their own institutional status, give implicit stamps of approval or disapproval to particular visions or versions of science. That is, they may act as authors with special authority on science—as authorisers of science.[28]

American museums in the 19th century were primarily educational institutions, and the American Association of Museums has reiterated its commitment to using museums for education.[29] For many museums still, and for most in the past, education means a classroom and a teacher, and something very like a school experience for the visitor. More recently, museums have embraced the idea of informal education, which involves less prescriptive, almost incidental learning through play and interaction of the kind embodied in science centers. But in any type of museum, while the curriculum may form the bedrock of many museum activities and exhibitions, the visitor chooses which lessons, if any, to learn, and how to learn them. According to Bloom:

> A museum visit is self-structured. There are no performance standards. No one can fail a museum visit. Our visitors are free to explore without fear of failure; to return again and again to something that interests them. I believe absolutely that the museum is a place of learning. It is a special place where people can follow their own interests; browse until they find something that inspires more focused attention and perhaps lights a spark that burns for a lifetime.[30]

One criticism that is made of science centers is that they take scientific principles out of both the natural and social worlds. Experiments or demonstrations with apparatus become instead games with toys; and the scientific ideas involved are displayed as the way the world is, rather than the way in which a particular community understands the world to be. According to Butler:

> What science centres do not generally make clear is that the demon-
> strations they present to the public are part of an existing knowledge
> system. There is a danger that science is presented as simplistic truth, a
> mirror image of a "real" physical world.[31]

Butler suggests that it is arrogant and naive of science centers to presume that their visitors would be unable to comprehend their experience within a particular conceptual framework, and this is a charge that has also been made against more traditional science museums. Science centers have also been accused of neglecting the social and historical context of their exhibits. Made from contemporary materials, not even the objects themselves give any clues as to their origin in history and society. Here science centers are at a disadvantage compared to more traditional museums, in which, with both the material of science and historical scholarship at hand, the curators have more opportunities to present their exhibits in context.

On the other hand, traditional science museums seem to score less well than science centers when it comes to communicating the ideas of science. As Butler points out, science is essentially a system of ideas, and museums are generally about objects.[32] Richard Gregory of the Bristol Exploratory has often claimed on the same grounds that there is actually very little science in science museums.

One attempt to deal with this line of criticism—as well as to make better use of buildings and facilities—involved the Earth Galleries of the London Natural History Museum. Up to 1994, the main hall of the museum contained dazzling collections of gems, minerals, and fossils displayed for public admiration. Attempts to explain the geology behind the formation of the exhibits were confined to side galleries—although one of these had to be traversed to reach the popular earthquake simulator. Reports commissioned for the museum indicated that, having sampled the exhibits on the ground floor, few visitors made it to the first floor, where British fossils were arranged according to their geological time sequence; and fewer still—only the dedicated "specialist" visitors—climbed the next flight of stairs to visit the collections of the British Geological Survey.[33]

During 1995, the Earth Galleries were—quite literally—turned upside down, and the emphasis shifted to demonstrating geological processes. Visitors to the new galleries, which opened in 1996, now make their way through an impressive entrance hall in which the processes at work in our Earth are represented by figures from mythology: Medusa, for example, towers over a specimen of a large ammonite, representing the process of fossilization as turning living creatures to stone. From the entrance hall, people are transported to the top of the museum via an escalator that passes through the center of an artistic representation of the Earth, wrought from iron and 10 meters in diameter. On the top floor, items from the old collections are used alongside interactive exhibits to illustrate the process of land building, continental drift, and erosion, with the earthquake simulator as the apex of the displays. A radical departure from the old collections-based approach, the new galleries have not been without their own difficulties, however. At a practical level, the wrought-iron Earth, which used to rotate as the escalator passed through it, developed mechanical problems that halted it. Many of the old specimens are being displayed once again. And a resource center is being provided separately to cater for the Galleries' more specialist visitors.[34]

If science centers discard the context and display the principles, and science museums hide the principles deep within historical objects, maybe combining the two types of institution might be a productive move. While the example of London's Earth Galleries illustrates that the new thinking is by no means without its problems, in the late 20th century the assets of both science center and museum have nonetheless joined forces, often after heated curatorial debate, in many institutions around the world. But whether any of these strategies has any effect on the visitor is an open question.

## EVALUATING THE VISITOR'S EXPERIENCE

At the Centennial Exhibition in Philadelphia in 1876, the centerpiece was two enormous steam engines, which, according to Bruce, towered "over the awestruck crowd like shining Buddhas endowed with muscle and motion."[35] What the visitors to this particular exhibition did not notice, however, were two very recent innovations that would shortly transform American lives: the electric light bulb and the telephone. Such technology, it was claimed, was simply beyond the public's imaginings: the *Journal of the Franklin Institute* lamented that while the Great Exhibition of 1851 was comprehensible to the average visitor, the intervening 25 years had transported science out of the grasp of laypeople, leaving them only to wonder at marvels.

Wondering at marvels must be a popular pastime: estimates suggest that there are between 25,000 and 35,000 museums worldwide,[36] and science museums are a long-standing and growing part of that total. While many museums subscribe to notions of the intrinsic value of furthering the public understanding of science (indeed, many science museums have been key centers of public-understanding-of-science activity and research), more and more commonly they charge an entry fee. The visitors have become customers, and museums are increasingly responsive to their needs. But what these needs are have proved difficult to pinpoint. Visitor numbers are not large when compared to other media: as museums consultant Roger Miles points out, even major national museums—the Air and Space Museum at the Smithsonian Institution, or London's Natural History Museum, for example—can claim only as many visitors in a whole year as watch a single edition of "NOVA" or "Horizon" on television in their respective countries.[37] Nevertheless, museums continue to attract visitors and enjoy a strong social mandate. So why do people visit museums?

Prime among claims for the attraction of museums is that they offer a close encounter with the real thing, be it a shark in an aquarium, some moon rock in a space museum, or a steam train in a heritage center. As Bloom confirms:

> A museum offers direct, one-to-one personal experience, the chance to experience real objects. Even in this media saturated "information age," there is no substitute for the power of reality.[38]

Another theory about what makes museums attractive is that people can construct their own experience of them as a medium. Unlike a television program, which goes at a certain speed and tells a story in a certain order, a museum exhibition can be wandered into and out of at whatever pace and in any direction.

One casualty of this is narrative: the kind of stories that might, in other media, be told about science—the development of a theory, or the progress of a discovery—are reduced to a collection of isolated episodes of ideas or history. In an exhibition about evolutionary theory or the history of aviation, visitors will collect their own selection of images and data and put together their own story, irrespective of how adamantly an optimistic curator may have designed an exhibition to encourage the visitor to take a particular route. Curators learn to accept that visitors have a mind of their own. According to science center director John Beetlestone and colleagues:

> A high degree of customer autonomy is retained throughout the activity so that the nature of the experience and any outcomes are determined as much by the participants as by the designer of the activity.[39]

Beetlestone and colleagues claim that the unstructuredness of the museum experience is empowering for visitors: they themselves, in the choices they make about how to structure their visit, are making decisions about science:

> Most visitors are intimidated by science. That's why science centres
> exist. Yet everything in a science centre is, by definition, scientific.
> Within minutes of entering the centre, visitors are making decisions
> about overtly scientific experiences: "Oh, that one looks interesting,"
> "I want a go on this one," "Look at these." The decisions are small and
> arbitrary, but they are about scientific apparatus; and the quality of de-
> cision, and the level of information that informs the decision, grow as
> the engagement grows.[40]

Visitors to museums are people who have chosen to step through the door—
or who have a parent or teacher who has chosen to take them there. They tend to
be from among the better educated members of the public, though school groups
encompass a broader range of visitors; and it may be children, inspired by a
school trip, who introduce their parents to the science museum. But despite being
used to learning experiences, most visitors do not behave like attentive pupils:
they "skim" the museum, just as they might glance through the pages of a newspaper, picking and choosing stories about the world. Museum visitors tend to keep
moving, and if and when they stop to have a closer look at something, they will
stop for an average of less than 30 seconds. Indeed, visitors' behavior in galleries
seems to indicate that the most interesting attraction is the exit.[41] Visitors get very
tired in museums, which is perhaps why they will spend at least half of their time
in the gift shop, restrooms, and café. Researchers have suggested that this "museum fatigue" is due not to the visitors' constant locomotion, but to the intense
bombardment of information that assails them as they wander through the halls of
science.[42] Museums researcher Gaynor Kavanagh suggests that people who
emerge bloodied but unbowed at the end of a day in a science museum often experience the satisfaction of an obstacle conquered rather than of a visit enjoyed.[43]

How and what people actually gain from a visit to a museum are extremely
difficult to measure. The "visitor studies" community of researchers is struggling
with all the usual difficulties of social science fieldwork, as well as with finding
the staff resources needed to muster a decent sample size, and with the ethical
problems of invading visitors' privacy and of interfering with what may be someone's one and only visit ever to a science museum. If visitors do make up their
visit as they go along, and interpret the museum from their personal perspective, it
is not surprising that researchers find it difficult to frame their research. Another
reason why the museum experience is so difficult to measure is that museums
nowadays tend to be multimedia institutions: video, interactive exhibits, objects,
sound tapes, computers, and text vie for the visitors' attention. Given the problems
media researchers face when assessing the effect on the public of a single

medium, such as television or newspapers, it is hardly surprising that a museum presents an enormous challenge to effects researchers. As in other effects research into other media, what visitors take away, affectively and cognitively, from a museum visit is very difficult to evaluate. In the opinion of Beetlestone and colleagues,

> Few, if any, evaluations have led to conclusions that stand up to rigorous methodological criticism. Fewer still have produced conclusions that are applicable beyond the bounds of a specific activity in a particular location.[44]

In particular, as Beetlestone et al. point out, research conducted on the premise that the museum is a place of learning tends to emphasize cognitive gains over any other. These gains—facts remembered, theories understood—are also much easier to measure than affective gains—changes in attitude, kindled interests—which are often personal and particular to the individual visitor, and which may not manifest themselves for a very long time.[45] As conclusions emerge from this very young and challenging field of research, the news is at first sight depressing for the educators: people learn very little by way of facts or theories or information during a visit to a science museum. It would be easy enough to construct a "deficit model" of museum visiting which would demonstrate that most visitors noticed hardly any of the exhibits and left having acquired no new information about science.

So how does a museum reach its intrepid explorer or weary meanderer? Again, research is difficult, and the results are tentative. Interactive exhibits, such as those found in science centers, seem to catch the visitor's eye and hold their attention more effectively than static objects. But for imparting information, objects, interactive and otherwise, tend to be rather reticent: as Roger Miles and Alan Tout of London's Natural History Museum point out, "as far as lay visitors are concerned, the non-verbal language of real things is no more than a museological conceit."[46] When people want information in museums, it seems they want text: clear, concise, meaningful text to be sure, but words on walls are where people look for the facts. But facts are easily lost, and they are also, it would seem, mostly, for most visitors, beside the point: research indicates that a museum visit is primarily an affective and motivating experience, and not about knowledge at all.[47]

Museums may awaken an interest, or stimulate further engagement with science in other forums.[48] There is much anecdotal evidence from visitors that a museum experience has been the key that turned scientific indifference into scientific enthusiasm; the experiences of people who work in museums, and who engage with visitors, would seem to reinforce this. According to Alan Friedman of the New York Hall of Science,

> . . . a small but crucial amount of learning takes place, with visitors getting excited about perhaps one thing out of 150 presented. That small piece of knowledge can, however, produce a large affective change in the visitor. Museums can awaken a desire for education through this small piece of learning or interest.[49]

The very existence of visitor studies research indicates a shift in emphasis in the museum community—one that has caused some friction in some institutions. The shift is from simply conserving culture to also communicating it; from advocating science to debating it; from the museum as a bank vault of the history of science to the museum as a living treasure trove of the scientific culture in which all can share. This emphasis on the public context in which their efforts are judged is a change that today's museums ignore at their peril.


## THE POLITICS OF MUSEUM SCIENCE

This chapter ends where it began: at the Smithsonian Institution in Washington, D.C. The Smithsonian was founded in 1846 "for the increase and diffusion of knowledge among men" and is now a large organization of many parts. Alongside libraries, research facilities, and a publishing house are museums of natural history, air and space, art, sculpture, and American history. It was in the Museum of American History in 1994 that a new permanent exhibition, *Science in American Life*, provoked a heated controversy and had many political repercussions, both within and without the Smithsonian. The exhibition was funded largely by the American Chemical Society, which was keen to support a historical exhibition that would make good use of computer interactive exhibits and that would interest children, and in particular girls and children from ethnic minorities. However, this patronage and these ambitions did not necessarily mean that the exhibition would present a rosy picture of science. On the contrary: *Science in American Life* prompted outrage in the scientific community—a letter of protest from the American Physical Society claimed that the exhibition was "seriously misleading" and would "seriously inhibit the American public's ability to make informed decisions on the future of science and technology."[50] Some changes were subsequently made to the exhibition, but it remains, largely intact, at the Smithsonian and is proving popular with visitors. The debate over *Science in American Life* was complex academically and politically, and painful and difficult for many of the people involved. The story merits closer examination than we will give it here. But it offers, in the context of the public understanding of science, a useful focus on many of the issues facing science museums as they negotiate their relationships with government, community, and science.

In many ways, *Science in American Life* is not an extraordinary exhibition. It includes hundreds of objects, plenty of text, and some multimedia computers. There is a "hands-on" gallery with interactive exhibits. Five themes, in content and in time, are explicitly identified to the visitor. The first, "Laboratory Science Comes to America," is set around the end of the 19th century and includes a replica of a lab in which chemists Ira Remsen and Constantin Fahlberg are discussing their discovery of saccharin. The next theme is "Science for Progress," which is exhibited through the artifacts and images of the 1939–1940 World's Fair in New York. The third theme is "Mobilizing Science for War" and deals with the atomic bomb, behavioral psychology, and the mass production of penicillin. The next is "Better than Nature," an exhibition of the synthetic materials that invaded American homes in the 1950s, the contraceptive pill, and insecticides such as DDT. The final theme is "Science in the Public Eye," which deals with areas of contemporary science that have made the news: ozone depletion, gene technology, and the superconducting supercollider. An underlying theme running through all the galleries is the underrepresentation of women and minorities in American science.

Yet *Science in American Life* dismayed some sectors of the scientific establishment and was debated in the U.S. press. A more detailed look at the content of the galleries not only provides some clues as to why, but also exemplifies the multiplicity of interpretations that can be made of museum exhibitions. Two reviews of the exhibition, one by Alan Friedman of the New York Hall of Science, and the other by Robert Park of the American Physical Society, offer contrasting views. For example, in a laboratory scene that includes life-size models of Remsen and Fahlberg, a tape recording plays a conversation between the two scientists about Fahlberg's incidental discovery of saccharin while working in Remsen's lab. Remsen thinks Fahlberg, who is interested in marketing his discovery, is bent on personal glory and financial gain and that he is giving too little credit to his colleagues. Later, Remsen gets involved, at President Theodore Roosevelt's request, in a debate with the food industry about food adulteration. In Friedman's account, Remsen and Fahlberg's conversation is "a lively debate," and the exhibition shows that "scientists are social beings, often engaged with the culture around them and sensitive to the impacts of their work on both science and society

. . . [they] are not always able to predict how their work, or their personal feelings, will play out."[51] According to Robert Park, the taped conversation is "a bitter dispute," and the message of the exhibition is that "scientists are vulnerable to vanity and greed."[52] A text panel alerts the visitor to the fact that all the people depicted alongside Remsen and Fahlberg are white men. Park is incensed:

> In 1890, you may be tempted to protest, the same would be true of stockbrokers, or law students, or architects—or historians, for that matter. But that seems to be the point; science, the exhibit is trying to tell us, merely reflects the values of the culture in which it is immersed and has no greater claim to authority than any other activity.
> And this is just the first stop. There is more than a century of this politically correct, postmodern social constructivism still to come.[52]

Park's last remark places this debate in a wider context: *Science in American Life* opened at the height of the "science wars," in which scientists and sociologists fought over the authority of scientific knowledge. As we outlined in Chapter 3, this debate was strident and sometimes bitter and often left little room for anything but extreme positions. In that environment, Park and Friedman's differences of opinion on the exhibition's treatment of science in wartime seem relatively mild. According to Park:

> . . . "the Bomb" . . . dominates the huge exhibit in terms of space, technical sophistication, and emotional impact. . . . the mushroom clouds, the scarred victims of Hiroshima . . . . The Cold War nightmare is dramatically captured with a complete family fallout shelter transported intact from the fifties. The atomic bomb is the ultimate symbol of science as an instrument of death and terror. What could an exhibit possibly offer to "balance" nuclear weapons? . . . Perhaps the development of antibiotics . . . but the meagre space devoted to penicillin is taken up largely with the development of the biochemical mass-manufacturing process—there is barely a mention of the impact of antibiotics on human health. No hint that penicillin . . . has saved many thousands of lives for every life taken by nuclear weapons and radioactive waste.[53]

Friedman, however, thinks that the exhibition shows that in wartime:

> . . . scientists responded with commitment and accomplishment. . . . The pharmaceutical industry learned in a few years how to ramp up the manufacture of penicillin . . . . Dropping the bomb on Hiroshima and Nagasaki is treated with extreme caution. . . . Sticking to its major theme of how the attitudes of Americans changed towards the role of science in their lives, the exhibition focuses on a home fall-out shelter in America, rather than on the direct consequences of the nuclear

bombing in Japan. A complete home shelter, retrieved from beneath a middle-American lawn, evokes strong memories . . .[54]

Of the exhibition overall, Park concludes:

> The message, delivered over and over, is that Western civilization is heavily burdened with guilt, and science, as a servant of a power structure, must bear a large share of that guilt.[55]

Park also commented, in the *Washington Post*, that "no opportunity has been lost to link science to social injustice."[56] Friedman responded thus:

> I find it hard to dispute that the public's attitude towards science in American life is far more cautious today than the earlier, near-Utopian view of a science-based society portrayed at the pavilions of the 1939 World's Fair. . . . Whether or not more positive case studies had been included [in *Science in American Life*], however, the exhibition would still have the burden of interpreting how American public opinion developed into today's respect for the achievements of science combined with scepticism for everything offered as a "technological fix" to a social need.[57]

In offering an account of science in American life, then, the exhibition has shown, according to Friedman's analysis, how public ambivalence to science arose in the United States; and in order to do this, it had to portray the rough with the smooth. The exhibition's remit is explicit in its title: it deals with the extrascientific impact of science—that is, it deals with the social, rather than with science itself. Park wanted a rather different exhibition:

> . . . it is science that uncovers the problems, and it is to science that we turn to put it right. Not because individual scientists have any claim to greater intellect or virtue, but because the scientific method sorts out the truth from ideology, fraud, or mere foolishness. What people need to know, and are not told, is that we live in a rational universe governed by physical laws. It is possible to discover those laws and use them for the benefit of humankind.[58]

Friedman notes how unusual *Science in American Life* is in its treatment of science: unlike the celebratory world's fairs or the promotional exhibitions funded by corporations, it takes a critical look at science. Unlike curriculum-based exhibitions that deal in basic principles, it deals with history and context. Friedman attributes the strong reactions to *Science in American Life* at least partly to its unexpectedness in this regard. He also points out that the scientists concerned about the image of science the exhibition presents to the public have very little evidence of actually what image the public take away from their visit. Indeed, science communication scholar Bruce Lewenstein reports that the Smithsonian's own survey showed that visitors were very positive about science when they arrived and just as positive when they left; and even critics of the exhibition acknowledged that people have enjoyed it and have not felt negative about science after seeing it.[59] The hands-on gallery was, in these lengthy battles, entirely uncontentious, perhaps reflecting that when science is taken out of context, it becomes unarguable. It is in this gallery, at the end of the "prescribed route" through the exhibition, that visitors are asked to take part in an interactive survey that includes, among other things, questions about their attitude to science: most people—whatever they may have picked up from the previous exhibits—are still very positive toward science.

Lewenstein has analyzed the exhibition in the light of the controversy around it. He notes how an exhibition's design can determine its emphasis: how, for example, one side of a particular debate might be obscured by dim light or because the visitors' attention is distracted by something more eye-catching. In these as well as in many other practical ways, museums as a medium are complex and challenging communication tools. But Lewenstein's overall impression of the exhibition is that "it is not anti-science; moreover, the charges against it reflect an unwillingness on the part of both scientists and historians and sociologists of science to understand the value of the perspectives on science that each side brings."[60] He also points out that many people who are concerned about the public image of science are concerned precisely because the public's image is too good: the public have expectations of scientists that are simply too demanding and unrealistic. In that regard, *Science in American Life* is a useful lesson in the technical, political, financial, moral, and personal problems that can beset scientists.

Again, though, there may be scientists who regret that people's high expectations mean that science is occasionally found wanting in the eyes of the public, but who nevertheless do not want to step down off the pedestal just yet. Lewenstein concludes:

> In the *Science in American Life* exhibition, the dispute is between scientists wanting to celebrate their intellectual achievements and their role in American progress, and historians using a broader definition of "science" (to include culture, institutions and people) who wanted to understand the relationship of that broad definition to other trends in American society.
>
> To me, the choice is clear: No one perspective can or should prevail. History is precisely the telling of multiple stories about the past, that together help us to grasp the totality of what happened in the past.

... The challenge ... is to find ways of conveying that complexity to the public, to contribute to the public's understanding of science. *Science in American Life* may not be the best of all possible exhibits at presenting this complexity. But it is, I believe, a fair and honest portrayal of the role of science in American life, and is a worthy contribution to the overall debate. As such, ... it contributes to the public understanding of science. Those who object to the story it tells are, I think, shooting the messenger.[60]

Well aware of the difficulties of finding out the extent to which the exhibition impacts on the public understanding of science, Friedman speculates as follows:

I think visitors will come away pleasantly surprised that the scientific establishment is able to consider itself so thoughtfully, without the hard sell for the benefits of science which the establishment musters before congressional committees. Scientists recognize that their work may significantly affect daily life, and that they must be self-critical when they use science to try to improve the human condition. That is an image that could go a long way towards restoring hope and promise to the public image of the scientific enterprise.[61]

## Chapter 9

# Initiatives and Activities in the Public Understanding of Science

———————◆———————

**Much of the activity aimed at furthering the public understanding of science**
long predates the overarching theoretical and political frameworks in which these
efforts are now seen. In some ways, the public understanding of science movement has hitched a ride on the public's own efforts, as exemplified by the grass-
roots activity of hobby clubs and illness advocacy groups among many others, to
indulge their interests in and satisfy their need for science. But in recent years
public engagement in science has been facilitated and encouraged by powerful
agencies with various agendas for changing the relationship between science and
the public. Governments, the scientific community, and business interests have all
offered the public new opportunities to participate in scientific events, be they celebratory festivals or critical inquiries into matters of public policy.

Every activity in this area finds its own particular balance between information and education, on the one hand, and advocacy and persuasion, on the other.
Researchers popularizing their work tread a fine line between sharing their enthusiasm and lobbying for funding for their particular pet project. Industry-linked resources for schools can easily become a straightforward exercise in public
relations. And government information on, for example, safe sex can easily spill
over into the party politics of "family values." These tensions are not always apparent: the agendas at work are often tacit, and the public's perceptions of and reactions to them are difficult to access.

In this chapter we pick out just a few examples from the plethora of activities and initiatives in the public understanding of science to illustrate this dynamic and diverse area of cultural activity.[1] Insofar as this chapter discusses work done with school and college students, it is from a standpoint outside of formal science education—an area which has received ample attention elsewhere; we confine ourselves to the ways in which the scientific community and others have set out to augment or illustrate the science offered within the confines of the classroom.

## SCIENTIFIC COMMUNITY INITIATIVES

Many scientists work in publicly funded laboratories or in universities or specialist research institutions that rely on the generosity of taxpayers. In the private sector, public tolerance of research activities is usually considered a minimum requirement; and products must appeal to consumers. While, as we discussed in Chapter 2, there have always been those "visible scientists" fired with enthusiasm to take the message to the public, today's scientific stars are joined by thousands of others working on small science exhibitions, producing resource materials for schools, or giving talks to community groups. Such activity is often—consciously or inadvertently—multipurpose: while genuinely wishing to be informative or entertaining, scientists may also be popularizing science with the aim of promoting their area of research either to other, perhaps skeptical, scientists or in order to recruit students, and in the hope that funding may be maintained or even increased.

Many of the scientific "trade associations" have introduced schemes with aims both of supporting local initiatives and of developing national programs of public-understanding-of-science activity. Far and away the most active of these is the American Association for the Advancement of Science (AAAS). Although slightly younger than its British counterpart, the AAAS is a much more extensive (and wealthier) enterprise. It has its own weekly magazine, *Science*, which is a vehicle both for high-profile research and for the popularization of such work. With over 300 staff in 1997, the AAAS can carry out activities similar organizations in other countries can only dream of. The organization's web pages (http://www.aaas.org) offer hours of browsing. Some of the AAAS schemes to improve public understanding of science are wide-reaching and ambitious. Project 2061, for example, which is outlined in Chapter 1, proposes a reform of the U.S. educational system "to improve science and technology literacy for all citizens."[2] The AAAS has its own Committee on the Public Understanding of Science and

Technology (COPUS&T), whose goals include coordinating the activities of some 170 U.S. scientific societies to address science communication issues across the country. In Britain, the equivalent committee, COPUS (also discussed in Chapter 1), works at a somewhat less ambitious level. Since its foundation in 1985, however, COPUS has introduced several schemes to promote science to the public, including an annual science book prize, and has developed links with organizations such as the Women's Institute.

Initiatives from the major scientific organizations have been introduced in many other countries. In Canada, for example, a Women's Science Network has been running for many years. In 1986, Mexican scientists brought together journalists and broadcasters to form the Mexican Society for Popularizing Science and Technique. The Australian Academy of Science has, since the 1990s, taken an active role in raising funds to increase public awareness of science and technology. In 1997, the Commonwealth Scientific and Industrial Research Organization had, in its Corporate Affairs Division, a budget of several million Australian dollars for efforts to increase public understanding of science. In South Africa, emerging anew from years of white minority rule, the Foundation for Research Development is instituting urgent measures to improve the country's generally low level of scientific literacy.

As well as having input into formal science teaching, scientists have also been involving themselves in "value-added" projects within the education system and beyond. One of the largest of these is the U.S. National Science Foundation's "Informal Science Education" (ISE) program, whose twin aims are to "promote the discovery, integration, dissemination, and employment of new knowledge in service to society" and to "achieve excellence in U.S. science, mathematics, engineering and technology education at all levels."[3] For this to happen:

> The findings and methods of research [need to] be quickly and effectively communicated in a broader context and to an expanded audience .... NSF aims to engage researchers and educators in a joint effort to infuse education with the joy of discovery and to bring an awareness of the needs of the learning process to research, creating a rich environment for both.[4]

This is very much a timely project; for example, in the 25 years to 1997, the number of American science centers that had "informal science learning environments" grew from 17 to around 300.[5] ISE projects include multimillion dollar mass media broadcasts on public and commercial channels, museum grants, and cash for youth and community groups to promote scientific understanding. These can be highly innovative: one community scheme, based at Columbia College in

Chicago, involved teenagers from the city's South Side housing projects in creating their own "ballet" of environmental chemistry. ISE grants can also be made directly to research scientists.

In Britain, research councils are involved in schemes to put their graduate students into the classroom, to bring some of the excitement of front-line science to young people. Similar activity is taking place in other countries: the "Let's Talk Science" program, based at the University of Western Ontario in Canada, trains volunteer university students and professional scientists to guide school students in hands-on activities. The European nuclear research center, CERN, in Geneva, Switzerland, organizes visits for parties of school students, who spend several days exploring the labyrinthine facility. In 1997, it appointed a senior scientist to oversee efforts to promote public understanding of particle physics throughout Europe. On a less grandiose scale, university science departments and research laboratories have, for years, given limited access to the public during events such as open-house days. There are also a number of privately sponsored activities, such as the Jason Project (named after the leader of the Argonauts), founded in 1989 by Robert Ballard, discoverer of the wreck of the *Titanic*. This annual adventure scheme gives school students experiences of discovery, such as exploring Roman shipwrecks in the Mediterranean (1989), and direct access to scientific institutions, such as the volcano research centers and infrared telescopes in Hawaii (1995).

It is clear that resources have to be provided in order for scientists to play an active role in initiatives directed toward public understanding of science. So there are now a number of schemes providing funds directly to professional researchers for work related to public understanding of science. The AAAS draws heavily on the budgets of the U.S. government, through, for example, the National Science Foundation, and attracts much corporate sponsorship. In 1996, the National Science Foundation began making "Informal Science Education Supplements" of up to $50,000 to its research grant holders to promote public understanding of science outside of the formal science curriculum. One of the schemes funded in the pilot year was a television documentary called "Understanding the Silence," which charted the efforts of a New Mexico research group to work out how "SNooZe" genes make cells go into the quiescent state of stasis. Others included a traveling exhibition, *Tales of Toads and Taste*, about the way amphibians "taste" their environment through their skins, and an interactive computer program which allowed its participants to chart the uses plants from different climates make of the water available to them. Even if American researchers do not have funds specifically earmarked for activities directed toward public understanding of science, it is still expected that major projects, at least, will make every effort to popularize their findings. Much of this work is carried out by young researchers, whose consequent mastery of popularization techniques makes them extremely attractive to industrial and commercial enterprises— an example of public-understanding-of-science activity enhancing the value of scientists rather than detracting from their "real" work. Funding for British scientists for public-understanding-of-science activities has been available, to a greater or lesser extent, for a number of years: COPUS has provided small grants since 1986. This funding has since been augmented by research councils covering the physical sciences, which encourage their grant holders to use 0.5 percent of any award for public-understanding-of-science activity. The Particle Physics and Astronomy Research Council (PPARC) has its own award scheme for small projects and, in 1997, initiated a scheme that makes awards of up to £100,000 for nationally significant activity. These schemes are open to research scientists and members of the general public. Networks of schools have been among the significant beneficiaries of PPARC small awards, and the Scotland-based "Big Science Roadshow" and a collection of demonstrations to popularize particle physics were the winners of the 1997 national awards.

The 1996 National Science Board survey of "Science and Engineering Indicators" found that 55 percent of American families had their own computer, and nearly two-fifths of these home computers were networked; around 50 million U.S. citizens were therefore "wired" from home.[6] Four out of five college students had access to the Internet, and, in many towns and cities, public libraries and popular cyber cafés offered opportunities to surf the net. Thus, U.S. scientists have, not surprisingly, led efforts to make use of the growing availability of the worldwide web, enormously speeding up the rate at which the latest high-profile science can become public property. Astronomers and space scientists were among the first to recognize the potential of the web for this purpose: during the impact of Comet Shoemaker-Levy 9 on Jupiter in July 1994, within just hours of each explosion, the latest images had come straight from the telescopes onto the web pages set up by establishments such as NASA's Jet Propulsion Lab. By the end of the first year, those maintaining the pages reported more than 5 million "hits" by members of the public (including the media and scientists not directly involved in the measurements). All that was as nothing, however, compared with interest aroused by the Mars Pathfinder landing in July 1997. There were 500 million hits during the Pathfinder's first month on the Red Planet; the peak hit rate was in excess of 46 million per day. The meteorological community, too, has recognized the potential of the web for providing up-to-date information, and many forecasts now come with detailed explanations of the processes responsible for causing the recent or predicted weather patterns. Expert advice and information networks available to the public on the Internet, such as Profnet (profnet@vyne.com) in the United States and the ExpertNet (paul.clarke@cvcp.ac.uk) in Britain, are also developing in many countries.

Since the beginning of the 19th century, the British Association for the Advancement of Science has held an annual weeklong meeting during which leading research scientists have given public lectures aimed at making the latest research as widely known as possible. The BAAS holds its meeting in a different town each year, and the journalists who attend produce a concentrated burst of science coverage during the week. The AAAS meetings have followed a similar pattern and are now much larger than their British counterparts. In the 1990s, however, the scientific community has also taken to organizing particular days and weeks during which universities, research facilities, government ministers, and the media have all been expected to fly the flag for science. The first science week organized on a European scale took place in November 1993 and was supported by national governments and the European Union; in 1997, around one million people attended some 1600 special events during Britain's Science Engineering and Technology week (SET97).

Some activities involve taking science out to the people. During SET95, for example, various university departments in Bristol set up stalls, complete with demonstrations and short talks, in a local shopping mall, where they attracted enthusiastic and, the scientists reported, rather knowledgable crowds.[7] At another event, the Wellcome Trust's Ph.D. students discussed genetics with people at small exhibitions prepared in collaboration with the Medical Research Council and displayed at major railway stations.

The concept of special science weeks and days has now spread to the African continent. In South Africa, the Grahamstown Foundation, together with the Foundation for Research Development, inaugurated an annual SciFest in 1997; and 1998 was designated as the first "Year of Science and Technology."[8]

## GOVERNMENTAL INFORMATION AND "OUTREACH" PROGRAMS

There are two main problems in trying to compare government activities between nations. First, definitions of what counts as public-understanding-of-science activity, as distinct from general information services or formal education, vary widely. Second, what, in highly centralized and statist countries, such as the United Kingdom, may be delivered by the central government, in more federal and privatized countries, such as the United States, may be provided locally or by funding sources other than the general public purse. The greatest proportion of government activities to increase public understanding of science are the day-to-day briefings from ministries or departments and the specific public information and education campaigns. These are often carried out in association with university or research institute scientists who, ultimately, are the ones providing information for government officials. As new findings become available or issues are raised by scientists, government officials, or interested sections of the public, government departments are either keen or obliged to respond. So health departments in the United States and Western Europe ran major campaigns during the 1950s on the importance of vaccination; in the 1980s, AIDS was a key issue. In rural Africa and Asia, government information agencies have been concerned with emphasizing the need for clean water supplies and with relaying the latest information about the most appropriate farming techniques. This "free advice" is, however, not always free of political or business interests: for example, during the early years of the AIDS epidemic, there was concern that the U.S. National Institutes of Health, which put out much of the information to the public, was too closely allied with the major pharmaceutical companies.

As well as performing their routine, and not so routine, informative role, governments also sponsor particular initiatives or projects aimed at promoting wider appreciation of science and technology. In several countries, the profile of science has been further elevated by the creation of a Minister for Science, or the equivalent, whose activities include overall responsibility for public understanding and attitudes; in the United States, this role now falls to the vice president. There, Bruce Lewenstein identified no fewer than 44 separate federal projects that were ongoing in 1993 that he could classify as public-understanding-of-science activity, quite apart from the efforts of departments and agencies having a particular science-oriented mission (such as the National Science Foundation, NASA, and the Smithsonian Institution).[9] These projects ranged from "Eating Right—the Dietary Guidelines Way" and the "Urban Treehouse Program," from the Department of Agriculture, to *Breaking Through*, a video sponsored by the Department of Energy that was designed to show how women achieve excellence in all branches of science. The offices of the U.S. National Oceanic and Atmospheric Agency offer detailed weather reports and forecasts, updated at hourly intervals.

A survey of scientific culture conducted in 1994 reported that "over the past decade, science has become an explicit issue of concern across much of Europe" for economic and political reasons and, at least as importantly, in an attempt to find a common cultural dimension to the European Union.[10] In some EU countries, governments have adopted specific public-understanding-of-science measures to be run either by the national research councils (as in Spain) or by government departments (as in Britain). Since legislators form a key part of the public for science, many countries now have state-funded organizations aimed specifically at keeping their representatives up-to-date: an example is Britain's Parliamentary Office of Science and Technology.

As well as national governments, several supranational governmental agencies are involved in public-understanding-of-science projects. The United Nations Educational, Scientific and Cultural Organization (UNESCO) is involved in an international campaign to put in place "structures and activities to foster a scientific and technological literacy for all, in all countries of the world" by the year 2000.[11] Project 2000+, as the scheme is known, faces great difficulties in producing "global minimum standards" in all the participating countries, given their widely different social and political structures and their very uneven economic development. But the organization has been warned against "propagating inferior culture-orientated [science and technological literacy] and numeracy programs for Third World countries."[12] Improving women's access to science and technology is also a key aim of Project 2000+, although critics argue that too little attention has been devoted to this aspect.[13] Other international efforts include meetings such as the "Earth Summit" conferences held in Rio de Janeiro in 1987 and in New York and Kyoto in 1997, in which governmental and nongovernmental groups aimed at reducing the destructive impact of human activity on the world's environment and climate.

## INDUSTRY

Other key players in providing scientific information to the public are industrial and commercial enterprises. One obvious motivation for the private sector to become involved in projects directed toward public understanding of science is the opportunity to overlap the general promotion of public scientific understanding and good public relations; for science-based industries, ensuring a continued supply of trained personnel is another. But an international review published in 1994 found that industry has been slow to respond to the challenge of furthering the public understanding of science.[14] To be sure, school visits and open-house days at factories and industrial installations have been commonplace in many Western countries for decades. Industry has also been an important provider of educational resources for schools and passes on technical information to a wider audience through promotional publications and reports. This may be done directly or through "trade associations," such as the American Chemical Society or Britain's Institute of Physics, or in "visitor centers" at industrial sites or research institutions—examples are discussed in Chapter 8.

Rather than initiating events or schemes, the private sector appears to see its role mainly as one of sponsoring the public-understanding-of-science initiatives of others. In the United States, for example, while the telecommunications giant AT&T and the pharmaceutical company Merck do produce large amounts of public-understanding-of-science material on their own account, many more high-tech companies prefer to fund museums and exhibitions or to support services to the media. Computer companies in the United States such as Intel and Hewlett-Packard have donated equipment for use in science literacy projects. British Telecom supports Britain's Science Line, a telephone information service that answers callers' questions about science. Many countries now have these telephone science lines, staffed by a mixture of paid and volunteer researchers who direct public (as opposed to media) inquiries to scientists when their own databases are unable to supply the requisite knowledge.

## LAY EXPERTS

If industry and commerce have been relatively slow to respond to a climate in which enhancing scientific knowledge among the lay public is seen as important, the same cannot be said for the pressure or advocacy groups whose interests often conflict with those of establishment science. Some members of these groups have become "lay experts"—formally untrained, but highly knowledgable. During the 1970s, American consumer protection groups successfully fought large corporations over the safety of their products. Global organizations with international memberships counted in the hundreds of thousands, such as Greenpeace and the World Wide Fund for Nature, have initiated and publicized important studies on species faced with possible extinction, using public outrage to force governments to bring in measures such as the near-total ban on whaling and the Conventions on International Trade in Endangered Species. Within individual countries, campaigns range from drives to prevent and clean up industrial pollution to making the public aware of threats to the natural heritage from overenthusiastic developers. Although many of these groups formed in opposition to modern science and technology, and to the relentless and destructive economic expansion they saw as the consequence, environmental groups have rapidly espoused science as helpful to their causes.[15] In Britain, for example, Friends of the Earth successfully commissioned its own geological studies to demonstrate that sites chosen as potential underground repositories for nuclear waste were unsafe.[16]

Medical interest organizations, too, have performed a vital role in educating their members, and the public at large, about the nature of the particular diseases they face. In the 1970s, feminists challenged many of the medical orthodoxies surrounding the understanding and treatment of "women's illnesses" such as breast cancer, and the relevant sufferer groups publicized their writings.[17] These groups became important sources of information for patients and doctors alike. In the 1980s, gay-community activists were not only able to influence the way in which people testing positive for HIV were treated, but they were also able to get right into the processes of medical research and drug testing and adapt them to their needs. The role played by these activists has now become a "paradigm" for other medical issue groups, according to sociologist Steven Epstein. That groups such as Act Up and the San Francisco-based Project Inform were able to exert such influence was due to four factors, says Epstein:

> First, activists imbibed and appropriated the languages and cultures of the biomedical sciences. By teaching themselves the vocabularies and conceptual schemes of virology, immunology and biostatistics, activists . . . succeeded in forcing credentialed experts to deal with their arguments.
>
> Second, activists . . . successfully established themselves as the voice of the clinical investigators' potential population of research subjects.
>
> Third, activists . . . gained credibility by yoking together moral (or political) arguments and methodological (or epistemological) arguments.
>
> Finally, activists . . . seized upon existing lines of cleavage within the biomedical establishment . . . sometimes tipping the balance.[18]

But such a close engagement with science brings its own problems: in becoming so entwined in the research process, many AIDS activists felt—and were viewed by their peers—as if they had "crossed over" and were representing medical science rather than their fellow sufferers. And, as in the case of the 1995 Greenpeace campaign over the dumping of the Brent Spar, which we discussed in Chapter 6, the use of scientific methods and research by people whose heart is in the right place does not necessarily make it "good" science.

One could argue that members of advocacy groups who undertake to educate themselves and their fellow members outside of the formal education system are building on a tradition of amateurism, in the best sense of that word. Although the AIDS campaign groups' input into research strategies, such as the organization of clinical trials, was vital, they did not carry out medical research themselves. Instead, they acted more as active intermediaries than as primary sources of knowledge, enabling their members and the interested public, as well as significant sections of the scientific community, to obtain the information they needed in a form suitable for them. However, there is still considerable scope for amateurs as researchers, armed with cheap instruments and, for certain subjects, access to the Internet.[19] Lewenstein identified 10 environmental organizations with U.S. memberships in excess of 100,000; in all, some 15 million Americans belong to such groups. Drawing on these resources, for example, every Christmas thousands of amateur ornithologists are mobilized by the Audubon Society to chart the voyages of migrating birds; these "twitchers" play a key part, too, in protecting rare species. In astronomy, thousands of amateurs organized into local associations play an irreplaceable role in looking for comets and studying variable stars. In geology, enthusiasts spend their leisure time finding fossils and exploring sections. Amateur natural historians—among them farmers, arborists, and anglers—are often the first to warn of species in danger or of the progress of new pests and diseases. And in archaeology, citizens equipped with metal detectors or just sharp eyes have made important discoveries. In some projects, such as the American "Earthwatch" scheme, members of the lay public are enrolled as volunteers in scientific research.[20] These amateur scientists have proved their worth even in the age of superconducting supercolliders and computer-controlled gene sequencing. Often, too, they are supported by and contribute to publications and broadcasts which give a sense of community and increase the effectiveness of their efforts. Popular campaigns to study the eclipse of the Sun may be coordinated through the pages of *Sky and Telescope* and *Astronomy Now* or through "The Sky at Night" television show; anglers have been mobilized to chart the progress of new parasitic diseases of salmon through the *Angling Times*.

## LINKING WITH AND THROUGH THE MEDIA

In Chapters 2 and 5, we outlined the past and present roles of the media in the public understanding of science; the overriding impression is that, for as long as there has been "science," there have been public outlets for its popularization. In the United States, readerships for magazines that concentrate on science and natural history add up to some 20 million; for two decades, many American mass circulation newspapers have had weekly science sections. What has changed, perhaps more in Europe than in the United States, in the 1980s and 1990s is the willingness of the scientific community to view the media as reliable and legitimate means of communicating with lay audiences. In the United States, the Scientists' Institute for Public Information (SIPI) has been organizing meetings between scientists and journalists since it was founded in the 1960s, with a view to breaking down the barriers between them.

Historically, scientific researchers have not been schooled in the techniques required for mass communication, so "media workshops," lasting from a day to a week, and designed specially for scientists at various stages of their careers, have been instituted in many countries. These have usually been organized through agencies such as the AAAS, SIPI, and COPUS or by research funding agencies, research laboratories (private and public), or individual higher-education institutions. International agencies, such as the Stockholm-based International Geosphere–Biosphere Program, have also shown an interest in courses for their research staff. In a typical short media course, working science journalists and broadcasters are brought in to give the scientists a "feel" for dealing with the media, getting the scientists to write short articles about their work, introducing them to the rudiments of radio and television science, or grilling them in mock press conferences. For those scientists interested in finding out more about life on the other side of the tape recorder or the camera, there are now opportunities to spend extended periods working with the media. The AAAS and COPUS have fellowship schemes that place working scientists with radio and television companies or with local and national newspapers, where they research, write, and broadcast or publish science stories. This account from a scientist who participated in one such scheme is typical:

> I enjoyed my fellowship enormously. I believe my work was appreciated, and I hope to put my experience to good use in the future . . . . I have learned a great deal about how to communicate science to a general audience. I think I am not bad at it, although the people I worked with at [the BBC World Service at] Bush House are orders of magnitude (with no apologies for the jargon!) better.[21]

Another former media fellow now criticizes as obstructive his fellow scientists who talk to journalists as if they were students with an examination to pass. These scientists simply fail to understand the pressure journalists are under, he says: "There is nothing quite like having to write three stories, on completely different topics about which you know absolutely nothing, within three or four hours."[22]

While there are now many schemes for helping scientists to understand journalists, there are fewer that try to overcome potential misunderstandings in the opposite direction. For many years, only the United States dared allow its science journalists through the hallowed portals of the research laboratory to experience life on the other side: a pioneer was the Marine Biology Laboratory at Woods Hole, Massachusetts, which runs a well-established Science Writing Fellowships Program; and some of the country's medical associations run shorter briefing workshops to enable journalists to keep up with the latest advances. In 1993, however, the European Initiative for Communicators of Science (EICOS) was started, taking 12 media professionals from 11 European countries (including the United Kingdom and Israel) to work in the Max Planck Institutes for Biochemistry and Psychiatry in Munich, Germany.[23] All of the participants spent at least one week on a program of laboratory exercises, demonstrations, and scientific lectures, to give them, too, a "feel" for how the other half lives. They also gave lectures to the institutes' staff on their experience of reporting science to the public in their own country. Seven of the journalists decided to stay to carry out extended laboratory projects. One participant commented that: "My main expectations were satisfied . . . I'll understand better the process of biological science: why to get a simple result is so complicated and long!" Another was similarly enthusiastic but made the telling point that the participants still saw a gap between themselves and professional researchers: "Journalists are hungry for information and want to know the state of the art. The working in the hands-on lab was a good experience. It gave a clear view of daily scientific work. But it did not make me feel a scientist; I remained a science journalist."

Of course, after such exchanges, journalists go back to being journalists and researchers to being researchers (though a minority have enjoyed their experiences so much that they have switched careers). In many countries, daily newspapers and broadcasting stations have long had their own specialist journalists and correspondents on the science beat. But one of the significant developments of recent years has been involvement of scientists more directly with the input of science into the mass media. The AAAS runs the web-based science news service EurekAlert (http://www.eurek.alert.org), with a range of stories for use by journalists. In some countries where the tradition of science journalism has been relatively weak—and particularly where regional traditions and distinctions are strong—the local science community has taken the initiative to get science stories and features into the media.

In the United States and Britain, professional media resource services have been established. In the United States, this grew out of the efforts of SIPI, although the organization no longer has responsibility for it, and the service is run by the research organization Sigma Xi;[24] in Britain, the pharmaceutical Novartis (formerly Ciba) Foundation finances the service.[25] Media resource services consist of a database of scientific experts—in the United States, a national list of over 30,000; and in Britain, a Europe-wide list, with some 7500 scientists—who are prepared to comment "off the record" to journalists who want to know more about a particular piece of science. Media inquiries come in by telephone, and the service staff check the database for a suitable source. The scientist is then contacted to see if he or she is willing to talk to the journalist or, if not, can suggest someone more suitable. The service is particularly valuable for journalists who do not specialize in science stories and may not have their own stable of sources as well as for journalists who are looking for a second opinion on some of the more controversial research they might wish to report, since, speaking "off the record," scientists not directly involved in the work in question are more likely to offer candid opinions as to its significance and reliability. The U.S. and British services are available over the Internet: the U.S. MediaResource is mediaresource@sigmaxi.org, and the UK Media Resource Service is mrs@novartisfound.org.uk. Other services are running in several countries in Europe, and the United Nations is sponsoring a similar project in Sri Lanka.[26] In South Africa, the Foundation for Research Development is setting up its own Science-Media Net as part of its activities during the 1998 "Year of Science and Technology." Adding to these specialist media services, universities and colleges increasingly publish media directories, listing their staff and their specialities. Individual groups of scientists, too—such as the UK Planetary Forum—advertise their members' expertise to interested journalists via the worldwide web.

As well as the general media science output produced independently by newspapers, magazines, and television and radio stations, there are several productions in which the media and scientists collaborate specifically. The American space agency NASA even has its own television channel, NASA Select, where journalists and aficionados can watch the latest Shuttle missions and catch up on the news from planetary missions. The AAAS ran workshops in conjunction with "Count on Me," a public television series aimed at teaching mathematics. Royal Institution lectures and science "masterclasses" have been televised on the BBC's "Learning Zone" by the Sussex University-based Vega Trust. On its own account, the AAAS has also been pioneering novel uses of radio to capture the imagination of the nation's 5- to 11-year-olds. It produces a regular program called "The Kinetic City Supercrew," which involves a team of youngsters who set out to solve the mysteries of the universe—such as why apples fall from the trees—in the manner of the Famous Five or Nancy Drew and the Hardy Boys. As a result of the show's great success—it has won a coveted Peabody Award for its innovative broadcasting style—television versions of the program may soon be seen on screens worldwide. During Britain's annual science week, several media have combined to take part in MegaLab experiments.[27] In one such experiment, readers of the *Daily Telegraph*, listeners to BBC Radio Four, and viewers of the BBC's "Tomorrow's World" science magazine program were asked to judge in which of two interviews a well-known broadcaster was telling the truth about which film was his all-time favorite. Listeners turned out to be the best judges, 73 percent getting the right answer; it appeared that television (52 percent right) gave too many confusing clues as to when the broadcaster was lying and when he was not, and the written word (64 percent right) too few. In another experiment, gardeners were enlisted to determine how far a particularly vicious, and horticulturally useless, carnivorous worm had spread throughout the country, at the expense of its vegetarian and soil-improving cousin. Other media initiatives in conjunction with scientists include devoting days or weekends to individual scientific stories or topics: in 1997, British television viewers got a "Weekend on Mars" to coincide with the successful arrival of the Mars Pathfinder probe.

## PUBLIC AND LOCAL COMMUNITY INVOLVEMENT

Alongside the growth in activity furthering public understanding of science is a growing band of critics who ask to what extent this enterprise really does allow the public access to science rather than simply invite them to celebrate it; whether what they get through this process is what they really need; and to what extent laypeople should be able to influence how science is produced and used. Once it is accepted that the public has rights and requirements in all these areas, there are different ways in which public rights may be realized and public needs met. Democratic governments can always claim, of course, that people can vote and that policies on scientific issues may be broadly decided upon through the normal democratic process. In reality, at least as far as national elections are concerned, science figures very low on the list of issues that move the electorate; at the regional and local levels, however, the environment, traffic or noise pollution, and industrial development may swing more votes. As we discussed in Chapter 2, concerns about the relationship between "ordinary people" and science, in all its manifestations, are by no means new. Moves to "open up" science have often followed periods of more general social questioning and upheaval: for example, after the 1968 Paris student riots and national strikes, democratizing initiatives in France tried to establish direct channels of communication between academic scientists and the public.[28]

Sociologist Alan Irwin has described the debate on the role of science in society as having been polarized between the Enlightenment vision of knowledge as a force for improvement and more recent highly critical accounts, such as those discussed in Chapter 3. Irwin asks: "Is it possible for . . . a citizen science . . . to emerge from these debates . . . ?"[29] He considers the "science shop" model of "science-citizen mediation" to be particularly appropriate in delivering to the public the information they need for their everyday dealings with science and technology. According to science shop pioneer Marcel van den Broeke, director of the PWT Foundation in Amsterdam, the aims of science shops are twofold: "to serve society by eliciting, identifying and meeting various groups' requirements for knowledge, advice and research results which may be provided by universities" and "to open up universities by making students and university staff more aware of practical issues in society which may be addressed through academic know-how and thus provide links between university research and the practical needs and questions of the non-academic community."[30] Science shops have been established, to a greater or lesser extent, in several countries: in the Netherlands, every major university runs at least one, and the annual budget for the scheme in 1993 was around $7 million.

Traditional methods of involving the public, in a limited way, in policy issues relating to science and technological developments have centered around special commissions and public inquiries. Legislators—themselves largely laypeople where scientific and technological matters are concerned—may set up committees of their own members or of eminent citizens to gather expert advice on issues as disparate as the use of human embryos in medical research and the consequences of discovering life on Mars. For high-tech installations such as nuclear power stations or industrial waste incinerators, the approval of planning applications is usually a matter of extreme public interest, and quasi-judicial inquiries are frequently held.[31] In many of these, interest groups and local communities have commissioned and presented their own studies on the feasibility, safety, and impact of the projects proposed.

More recently, a new mechanism for involving the public more directly has been developed: the "consensus conference,"[32] which puts science and technology "on trial" in a quasi-courtroom setting. During the 1970s, the United States Congress became concerned about the rapid expansion of medical services and treatments and their potential cost implications. Medical consensus conferences were set up to assess the new medical technology and to draw conclusions about its efficacy and about how far it should be made available. These conferences consisted of "trials" of the technology in which panels of medical experts and panels of care-givers, such as general practitioners and hospital staff, discussed the issues in front of audiences usually comprised of other interested health professionals. These medical technology assessment processes soon gained international acceptance. In the 1980s, a number of European countries, with Denmark and the Netherlands in the forefront, adapted the process by replacing the expert panels with members of the general public, in an attempt to make technology assessment more democratic. In Denmark, with its strong traditions of active citizenship, lay panels became the norm; in the Netherlands, using a more pluralistic concept of participatory democracy, the composition of panels more often reflected interest groups concerned with the issue under discussion. In both countries, the scope of consensus conferences was broadened to take in new technologies outside of the field of health and medicine, as well as issues relating to the environment and pollution.

In the Danish model of organizing and holding a consensus conference, the topic of the conference is decided by the organizing body (in Denmark, this is often the Board of Technology); and a small steering group, with a high level of expertise in the relevant discipline or disciplines, is established.[33] The lay panel is chosen from respondents to advertisements in the regional newspapers, and panel members are expected to attend at least two weekend briefing sessions to get acquainted with the subject of the conference. In addition to the lay panel, there is also a panel of expert witnesses upon whom the lay members can call, if they so choose, for information and opinions. During the preparatory weekends, lay members will often interview some or all of the experts and decide whom they wish to hear more from during the conference.

The conference per se typically lasts three days and is held in public, usually with the media in attendance. During the first two days, the experts called on by the lay panel make their presentations and are cross-examined. The panel then withdraws, to complete its own deliberations and draw up its report. On the third day, the panel presents its report at a public meeting and is questioned by the experts it has called and by members of the public audience.

The results of consensus conferences in Denmark, in particular, and the Netherlands have had important bearings on the outcome of the legislative process: as a result of lay panel reports, the Danish government decided, in 1987, not to fund research into animal gene technology projects and, in 1994, to make safety the top priority when introducing techniques from information technology into the management of traffic. Thus, the consensus conference is fairly well integrated into the decision-making processes of these two countries. Other European countries seem less receptive, preferring more traditional methods of sounding out the public's views, such as opinion surveys or simply the weight of letters for and against an issue in a member of parliament's or representative's mail. Fears have also been expressed that lay panels would be prejudiced against scientific and technological development—though experience seems to contradict this concern.[34]

The cost, in terms of time, money, and effort, of organizing a consensus conference can also be an effective deterrent, particularly when quick decisions are needed. Cheaper methods of involving the public may then involve small focus groups—much favored by the medical industry—where a discussion leader sounds out in depth the views of a few people, or deliberative conferences, in which experts debate technological issues in front of a relatively large audience of members of the public who then vote on the question. And there are other mechanisms for involving the public in decision-making about science, such as citizens' juries, where lay panels choose between a small number of preselected alternative strategies, and scenario workshops, in which, for example, residents' representatives, local authorities, and the business community meet to discuss plans for local initiatives.[35]

In 1997, Richard Sclove, a political scientist at Amherst College in Massachusetts, began a experiment called "Loka Alerts," which distributed by e-mail "essays . . . concerned with democratizing science and technology." According to Sclove:

> Scientific leaders are predictably ecstatic at the latest evidence that what they do is vital to American industrial performance. . . . But the rest of us should think twice before agreeing to pour more tax dollars into a science system that's become coupled much more tightly to business than to civil society, democracy or the broader public good."[36]

The impact of Loka Alerts in involving Internet users in the policy-making process remains to be seen, but this experiment may well herald a new era of electronic democracy—at least for computer-literate citizens.

# RESEARCH AND EDUCATION
# IN PUBLIC UNDERSTANDING OF SCIENCE

Historians, sociologists, political scientists, psychologists, educators, and some scientists carry out research into the public understanding of science and science literacy, and articles of interest are to be found in the research literature for which they write and to which they subscribe. In the 1990s, however, those who specifically identified research into public understanding of science as one of their major interests began to define a field for themselves. There are now two peer-reviewed journals: *Public Understanding of Science* and *Science Communication*. The topics covered in these journals range, for example, from fairly abstract inquiries into the use of rhetorical analysis to assess the reliability of scientific journals, through case studies of the media representation of the risks of new technology, to the efficacy of national science weeks in stirring the interest of shoppers in a local mall. Researchers can now exchange views and information through an electronic bulletin board, run from Cornell University (PCST-L@cornell.edu). A number of national and international conferences on the relations between the public, science, and the media have also been organized.

The importance of putting research into public understanding of science on a firm, rather than anecdotal, footing is being increasingly recognized by funding agencies. In the United States, much research in science and technology studies, including science communication, is funded through the National Science Foundation's Directorate of Social, Behavioral and Economic Sciences. Britain's Economic and Social Research Council has made a number of "special calls" for projects, and the European Union has a particular interest in funding survey research into public attitudes to science.

Britain's COPUS commissioned a survey of research published in the decade to 1996, which identified around 500 papers wholly or mainly relevant to public understanding of science.[37] The survey, carried out by science communication researcher Jon Turney, found that researchers were trying to find reliable answers to a wide range of complex questions. Among many Turney identified were:

- Does increasing public knowledge of science correlate with increasing acceptance and appreciation of it?
- Whom does the public trust to provide reliable information about health, safety, and technology matters?
- How large do risks have to be before they become "unacceptable"?
- Is the information provided by the scientific community the information that is needed?
- Are women's attitudes to science different from men's?
- How do schoolchildren's perceptions of scientists affect the way they learn about the subject?
- Do audiences for popular science enter the "learning process" with a clean slate, or do they learn from, and enjoy, having their existing notions challenged and falsified?

Since there are almost as many paths into public-understanding-of-science research as there are practitioners, it is not surprising that the research techniques used reflect the influences of many disciplines: historical surveys of public understanding of science make use of general historiographical methods; knowledge and attitude surveys are statistically analyzed; and case studies make use of well-established sociological techniques. Turney found that different researchers were bringing to bear different techniques to try to answer the same, or at least similar, questions. For example, two approaches were used by researchers trying to assess the extent to which levels of understanding of science affected attitudes toward it. In his study of public attitudes to biomedicine, Jon D. Miller used a randomized opinion and knowledge survey that included a number of open questions to give respondents more room to air their views.[38] John Doble and Jean Johnson, on the other hand, used a 100-strong "citizens' review panel," whose members filled in questionnaires prior to and after watching and discussing videos on issues such as global warming and waste disposal.[39] Reassuringly, from a methodological point of view at least, both approaches found that increased knowledge did affect the opinions expressed on the issues, but neither study found that more knowledge automatically led to a more favorable attitude to science.

During the last decade, the availability of courses on public understanding of science for university and college students has mirrored the growth of research activity in this area. To a first approximation, these courses may be divided into two groups: those based mainly on science communication, with a practical, often specifically journalistic bent; and those with a more historical, sociological, and/or philosophical approach. Schools of journalism throughout the United States and Europe deal with science communication as part of their degrees or diplomas. In the United States, Sharon Dunwoody, Elizabeth Crane, and Bonnie Brown have compiled a directory of between 40 and 50 universities that give courses specifically involving science writing.[40] In American universities in agricultural states, science journalism courses have grown out of the tradition of providing scientific information to local farmers; now, courses cover both practical and theoretical aspects of media science. In Britain, students of journalism may take courses in science journalism as part of their overall degree; undergraduate scientists may be encouraged to develop their communication skills by taking courses in science writing. But there is also a growing number of graduate programs that concentrate specifically on science communication. New York University, for example, offers a postgraduate program in science and environmental journalism, and Imperial College, London, established Britain's first master's program in science communication in 1991.

Communication researcher Jane Randall Crockett has highlighted the growing importance of professional training in "informal science education" in the armory of public-understanding-of-science resources.[41] Her survey of American informal learning centers found that most workers felt that "on-the-job training" was more appropriate than formal education in preparing them for dealing with the public. Nonetheless, some of Crockett's respondents, such as Alan Friedman of the New York Hall of Sciences, felt that a more academic approach might be beneficial, and there are now courses, such as that offered at the University of Glamorgan in conjunction with the Techniquest Science Center, that aim specifically to train science center demonstrators.

Less practical, more critical courses come under the academic heading of STS, which stands either for Science and Technology Studies or Science, Technology, and Society and means much the same in both cases. STS programs that examine the public dimensions of science have existed for many years in universities across the United States and have developed over the last 20 or so years in Britain. Many STS courses have grown out of humanities or science/engineering programs concerned with the social impact of science and technology or out of history or philosophy of science programs.

As the teaching of courses on science communication and public understanding of science has increased, so have the opportunities for reinventing the wheel. However, it is always a good thing if, once a reasonably round wheel has been invented, this fact is communicated as rapidly as possible. Multi-institutional seminar groups and workshops have been set up, in part to try and address this question. For the past two decades, science communication teachers in the United States have been able to share information through the *SCIphers* newsletter, published by the Association for Education in Journalism and Mass Communication. In Britain, a Science Communication Teachers network was established in 1994, with a conference held under the auspices of the Wellcome Trust, and extended to a European network the following year.[42] Public understanding of science is relatively young compared with other pedagogical enterprises, and debates about curricula and aims are still in progress.


# ENDNOTE

Because public understanding of science draws on so many traditions, there is an almost infinite scope for theoretical and practical activity. One of the problems is that too often the activities are widely separated from the theory: the practitioners come mainly from the community of active or retired scientists or are professional journalists or broadcasters, while those engaged in researching their activity and its consequences tend to come from the social sciences and humanities. Too often, too, the activist is unaware that anyone has tried to reflect on what has been done, while the researcher is interested only in recording and criticizing. One reason why activists sometimes behave as though they subscribe to the deficit model of public understanding is that this model describes a simple problem that is easy to solve: their task is to fill the presumed gap in the knowledge of the people who attend their talk, read their article, or tune in to their broadcast. Researchers, on the other hand, are increasingly lending their weight to the conclusion that what is being addressed by science in the public sphere is less like a vacuum and more like a complicated synergy of deeply embedded cultural themes and awkward issues such as trust and belief. This picture is much more difficult to specify, and its needs are not easily satisfied. The extent to which the lessons of a detailed study of public understanding of science can inform practical activity is the subject of the next chapter.

## Chapter 10

# A Protocol for Science Communication for the Public Understanding of Science

**Whatever the traditions of the past, and whatever the reality on the ground,** the fact of the matter is that we live in a time when the scientific establishment, and its governmental and commercial paymasters, are voicing concern about the relationship between science and the public. The call is for improvements in this relationship through an increase in what is variously described as approval, appreciation, or understanding of science by the public. In order to improve the public understanding of science, scientists should communicate with the public.

Although critics argue that what should be communicated, and why, is far from clear, the "public understanding of science movement" has charged into the arena of science in public. Alongside the long-standing public and amateur activity in science, there are now events, from festivals to consensus conferences, set up with the explicit aim of furthering the public understanding of science. While much of this originates with the scientific community and is not coy about aiming for the public appreciation of science, some activity arises from public groups or individuals who are dissatisfied with science.

But changing times have left a legacy of confusion about the role of the scientist in society, and in particular about the scientist's responsibilities in the public communication of science. Political and social circumstances have shaped the changing relationship between the scientist and the public, and different motivations at different times have impacted upon the content, style, and audience of science communication. New media, and new mediators, have complicated the picture in the post-war era.

The media's role in the public understanding of science has been much criticized by scientists. Examinations of the professional and social forces at work reveal a high degree of collaboration and mutual reliance between scientists and journalists, even though the differences between journalistic and scientific practices can sometimes lead to friction. The media do provide the forum in which the relationship between science and the public is constructed and pursued, and it is in this forum that the public make moral judgments about science. However, despite the media's activity in the communication of science, they have no brief or responsibility for improving the public understanding of science, and, in some cases, are ill-suited to the task. Whether media science has any impact on public knowledge about or attitudes to science is a question that research has yet to answer.

Detailed examination of particular cases of media science reveals the many strategies and agendas that are at work in science in public. From 19th-century didactic entertainments have grown media campaigns in which science politics are played out in public. One thing is clear: in contemporary Western society, media science is often "science-in-the-making." While science *is* in-the-making, the "right answer" is unavailable to everyone, including the scientists, and anyone who claims to know what it is deserves public skepticism, if not distrust. In such cases, the scientific literacy enterprise, in terms of knowledge or understanding, is beside the point.

Science often meets the public in times of crisis. Their relationship is conducted fleetingly and acutely through mass media that emphasize emotion in place of what are often rather scarce "facts." And when scientists cannot agree on a solution to a scientific problem, it is not surprising that the public make use of solutions based on moral or emotional considerations in order to get on with their lives. The highly charged environment pushes everyone involved to extreme practical measures and to polarized points of view and often results in a breakdown of both trust and communication between political and scientific authorities and the publics they purport to serve. Studies of risk communication suggest that responses to risk situations are informed by many factors other than the simply scientific and that finding a place for a scientific point of view may be achieved more through negotiation within social systems than by pronouncements of facts.

Science has long been subject to scrutiny from many different quarters. In recent years, critiques of science emerging from the social sciences and humanities, and the apparently growing popularity of alternative sciences, have been construed by some scientists as an "anti-science movement." This movement, they feel, threatens to undermine science and to deceive the public into adopting false and dangerous beliefs. This may be true of some critiques, which claim to offer an all-embracing alternative to scientific knowledge. Most, however, actually offer understandings of science that look at both laboratory practice and social, personal, and political contexts—the contexts in which the public experiences science. And while these understandings may cast doubt on the hypothetico-deductivist "standard account" of the scientific method, they make few if any charges against the reliability of scientific knowledge, nor do they question its value to modern society.

Much discussion of the public understanding of science rests on simplistic models of "public," "understanding," and "communication" and remains tacit on questions of the motivations for and gains to be had from the public communication of science. However, researchers have revealed complex mixtures of motives in scientists' communications, among which improving public understanding can play a minor and sometimes incidental part. Models of the public and their understanding have developed from a passive and empty "black box" to an active, discriminating body accommodating, should it so choose, information within its own cultural framework. Models of communication, too, have become more complex as the transmitter–receiver descriptions have failed to account for the variety of interactions observed between communicator and audience.

In this book, then, we have looked at the ideas of popularizers and at the workings of the media, and we have brought together the work of scientists, sociologists, historians, and philosophers who have all given a great deal of thought to the issues of public understanding of science. Now we look at how these people's ideas might be translated into an approach to science communication that meets the requirements of all concerned—scientists, communicators, and the public. What follows is our contribution to producing a protocol for communication to further the public understanding of science.

## ACKNOWLEDGING THE PLACE
## OF POPULARIZATION

Popularization is an essential and an integral part of the scientific enterprise. Scientists use popular accounts and popular media to reach each other, and even writing up a piece of research for a scientific journal involves a certain amount of "popularization"—of writing for people who do not know what they are going to read, and writing in a way that matches the readers' expectations and suits their background and sensibilities. Scientists take for granted that the scientific paper is not literally true: it is not a blow-by-blow account of what happened in the culture dish or under the telescope dome. But the scientific paper is truthful even though it is written according to a formula which deliberately distorts the literal truth in order to make the research accessible to other scientists. So popular accounts of science should not be viewed as somehow "untrue," merely because they, too, have to leave out a lot and simplify what they include to match the expectations and abilities of their audiences.

In some instances, science-in-the-making finds its way into the public arena. Even for uncontroversial subjects, part of the process of legitimation and acceptance of scientific knowledge may be the incorporation of the concepts and insights of science into everyday consciousness and parlance. The line between scientific knowledge and common sense is a historically moving one.[1]

There is no clear boundary between what is and is not "popular" science as opposed to any other science. Nor can the audiences for these communications be delineated exactly. The scientific community should therefore treat popular science as seriously as science in any other form. After all, if they do not, why should the public?

## BEING CLEAR ABOUT MOTIVES

Almost without exception, the best scientist-communicators feel, and are able to communicate, enthusiasm for the work they do and for the subject they love. But pure enthusiasm is rarely the only motive for communicating science. The purpose of a science communication may be to empower its recipients, to enhance existing democratic processes or help develop new ones where such processes do not exist, or to prevent the alienation of sections of society; but it may also be to serve the interests of the scientific community and its paymasters.

Scientists communicating science to the public should therefore make their motivations clear. If they do not, the public could usefully ask what their motivations are. The media have responsibilities here, too. Independent science journalists have a duty not to allow themselves to be used to promote one or other faction in a dispute over the facts, implications, or worth of a piece of science. If straightforward public relations is the purpose of the communication, then this too should be plain.

We have already referred to the benefits which might follow from increasing the public understanding of science, as collated by Thomas and Durant.[2] If the aim of the communication is to further the public understanding of science, then thinking about whether it affords any of these benefits might be fruitful.

People have their own reasons for paying attention to science. Collectively and as individuals, the public therefore have a responsibility to be explicit about their own motives for understanding science. If they are not, scientists may find it difficult to know what is expected of them.

## RESPECTING THE AUDIENCE

One of the purposes of science communication is to impart knowledge in a form that can be assimilated by the intended audience. This approach requires due consideration and respect for that audience. While they were writing *The Science of Life*, H. G. Wells gave Julian Huxley some good advice:

> The reader for whom you write is just as intelligent as you are but does not possess *your* store of knowledge, he is not to be offended by a recital in Technical language of things known to him (e.g. telling him the position of the heart and lungs and backbone).
>
> He is not a student preparing for an examination & *he does not want to be encumbered with technical terms*, his sense of literary form & his sense of humour is probably greater than yours.
>
> Shakespeare, Milton, Plato, Dickens, Meredith, T. H. Huxley, Darwin wrote for him. None of them are known to have talked of putting in "popular stuff" & "treating them to pretty bits" or alluded to matters as being "too complicated to discuss here." If they were, they didn't discuss them there and *that was the end of it.*[3]

Successful communication requires at least some grasp of technique: audiences expect and deserve competence. Few scientists are expert communicators, and many, finding themselves suddenly amateurs in the field of science communication, have usefully learned a few new skills for the job. There are now several agencies and professional organizations that provide short courses on science communication, and there are a number of useful books in this field.[4] Books aimed at journalists, or books that feature discussions on the relationship between scientists and journalists, can also give scientists and the public who use the mass media useful insights into the worlds of the popular press, television, and radio.[5]

## NEGOTIATING NEW KNOWLEDGE, UNDERSTANDING, AND ATTITUDES

According to the deficit model of public understanding of science, the scientific community is the source—and, by and large, the censor—of the information that is transmitted in a one-way stream to the public. The contextual approach, on the other hand, tries to take account of the particular circumstances of the recipients of scientific information and of their existing knowledge and beliefs. Sometimes—for a straightforward talk, article, or broadcast—the deficit model may well be the more appropriate. After all, most communication is about someone telling someone else something they did not already know, and the context may be unknown. But many case studies indicate that the straightforward "facts" of the matter do not always answer the questions that people tend to ask. No matter how straightforward the science, the recipient of the communication will be a complex human being whose background, beliefs, and sensibilities play a large part in their reactions to scientific knowledge.

Communication is a process of negotiation: it is one of a mutual getting-to-know. Science communication is a process of generating new, mutually acceptable knowledge, attitudes, and practices. It is a dynamic exchange, as disparate groups find a way of sharing a single message. Negotiation is a two-way process: if the public's needs are to be met, the public must articulate what these needs are.

## ESTABLISHING A BASIS FOR TRUST

If the stories and theories in this book hold any lesson above all others, it is that key to the relationship between science and the public is trust. This trust is established through the negotiation of a mutual understanding, rather than through statements of authority or of facts. Responsibility for the trust vested in the scientific community rests both with the institutions of science and with each individual member; it is hard won and easily lost.

Science has every right to defend its role as a provider of "reliable knowledge" in our society. Indeed, it has a duty to account for itself if it is to maintain the public's trust. This position does not imply or require that knowledge cannot be arrived at by other means, however. Western science is not the only resource for solving the world's problems, and respect for local culture and the assimilation of local knowledge are vital if the application of inappropriate or socially disruptive technologies is to be avoided.

Science is undoubtedly powerful—perhaps the most powerful of all forms of knowledge, in terms of its effect on our daily lives and social structures. Appreciating the power of science is not merely an interesting academic pastime, but an activity essential to the well-being of society. So, too, is appreciating the limitations of science, in terms of what questions it can and cannot usefully answer.

One of the key features of science is its inherent provisionality. When science is in-the-making, this provisionality is the *essential* feature of scientific knowledge. It is surely much better for scientists to acknowledge this so that the public and their representatives can make the best use of what information is available, from whatever source.

At various times, journalists have drawn up questions to ask scientists, designed to lay the basis for a genuinely trustworthy process of communication between them and the public:[6]

- How do you know? Are you just telling us something you believe? Or is it based on scientific research?
- How and where did you get your data?
- How accurate are your numbers?
- How reproducible are your results? Have they been consistent from study to study?
- What is your degree of certainty or uncertainty by accepted tests?
- How can you be sure about conclusions? Are there any possible flaws or problems in them?
- Who disagrees with you? And why?

These are questions that scientists ought to be prepared to answer, as honestly as they can, and that the public ought to be prepared to ask.

## ACKNOWLEDGING THE SOCIAL IN SCIENCE

Science is one of the major cultural achievements of our species, and it is right and proper that science should be subject to detailed social scrutiny, by academics and the public. Carrying out social, cultural, and historical critiques of science does not equate with being "anti-science." Scientists themselves should be part of this questioning process.

Analyzing the practices of science and scientists—how scientists conduct their arguments, how concepts develop within the limitations of individuals and the society in which they operate, how policy issues affect and are influenced by research—can not only be helpful in deepening general public understanding of science, but can also be beneficial to scientists wanting to know how they got where they are and where they might be going. Studies of social influences can usefully alert scientists to the possibility that "bad science" can result from racial, gender, and cultural prejudices.

Accepting that science may be criticized from various social standpoints in no way implies support for the position of "nothing-buttery"—the notion that science is "nothing but a socially constructed discourse." There are inevitably social influences on the work of scientists and on the development of scientific knowledge. But accepting this does not mean that such knowledge is somehow less valuable or less useful. Society has enabled humanity not only to survive as a species, but also to accumulate reliable knowledge of, and wisdom about, the world in which we live. Whether socially constructed or not, scientific knowledge mostly seems to work.

Like their colleagues in the natural sciences, social scientists have a duty to popularize their work, even when their conclusions are not quite to scientists' taste. Nor is it incumbent on those who criticize science from a social standpoint to produce an alternative. Neither theater critics nor literary reviewers are required to outperform their subjects, and actors and authors take note of their criticisms when they so choose. And it is not disputed that the public sphere is the proper place for both the performance and the criticism.

Scientists should be free to investigate the natural and social worlds as they see fit (within the laws of the land); they have a duty to communicate more than just the "bare facts" of science. As responsible citizens, they should be prepared to bring out the social implications of their own and their colleagues' work, voicing their optimism and enthusiasm where appropriate and their concerns and reservations where they have them. They are, surely, among the better placed to do this.

## FACILITATING PUBLIC PARTICIPATION

The business of science has, of necessity, become highly professionalized. But at the same time, the results of science have been more and more socialized in terms of their impact on the public, through changing beliefs, practices, and lifestyles. The public have a right to know about science not simply because they pay for much of it, but because of its central importance in the modern world. The public are interested in science: there is popular enthusiasm for knowing about facts and theories, and curiosity about how those facts and theories were derived.

Scientists thus have a duty to explain their work to the best of their ability—even when it requires the suspension of current common-sense thinking—and to be open about the potential, limitations, and practices of science. Scientists have a right to defend in public their ability to do research irrespective of the social uses to which that research has been or may be put. In the end, however, questions of which research will be supported, and how the results will be used, should—and will—be decided in the public sphere. This places considerable responsibilities on the public. If citizens are to fulfill these responsibilities, the communications they receive about science must enable them to participate in the social processes of debate and decision-making. Communications designed to bring about an awestruck admiration for the mysterious men in white coats are not what we need for the challenges of the 21st century.

Each of the territories explored here—science, the media, and the public sphere—represents for the inhabitants of the other two a largely unknown land. Like unworldly tourists, these groups are inclined to believe that if they speak their own language slowly and loudly, they will make themselves understood; sometimes, like imperialists in an annexed land, they presume that everyone else is a savage. Just as travelers abroad have learned to understand another culture on its own terms (and have, perhaps, even read the guidebook before they set out), so might scientists, journalists, and the public tread a little more lightly on each other's toes if they got acquainted first.

If this book has any message, it is that the entities involved in the public understanding of science are much more complicated than the movement's rhetoric implies. Simply acknowledging this complexity would be, we believe, a step forward; understanding that complexity, while a daunting task, has been made easier by the considerable body of work coming from the social sciences and from communications research, examples of which we have referred to in this book. Accessing these ideas could go some considerable way toward helping scientists understand the public, and the public to understand science.

# References

## CHAPTER 1

1. Martin Bauer, John Durant, Asdis Ragnarsdottir, and Annadis Rudolphsdottir, *Science and Technology in the British Press, 1946–1990*. London: The Science Museum: 1995.
2. John Dewey, The supreme intellectual obligation." *Science Education 18* (1934), pp. 1–4.
3. Weaver's statement is known as the Arden House statement, after the location of the meeting at which it was formulated. For the history of this episode, see Bruce V. Lewenstein, Public understanding of science in the United States after World War II, *Public Understanding of Science 1* (1992), pp. 45–68.
4. Warren Weaver, Science and people. In: *A Guide to Science Reading*, edited by Hilary J. Deason. New York: Signet, 1963; pp. 25–37.
5. S. B. Withey, Public opinion about science and the scientist. *Public Opinion Quarterly 23* (1959), pp. 382–388.
6. Jon D. Miller, Scientific literacy in the United States. In: *Communicating Science to the Public*, edited by D. Evered and M. O'Connor. Chichester: Wiley, 1987; pp. 14–19.
7. Morris Shamos, *The Myth of Scientific Literacy*. New Brunswick, New Jersey: Rutgers University Press, 1995.

8. Henry H. Bauer, *Scientific Literacy and the Myth of the Scientific Method.* Urbana: University of Illinois Press, 1994.

9. John R. Durant, Geoffrey A. Evans, and Geoffrey P. Thomas, The public understanding of science. *Nature 340* (1989), pp. 11–14.

10. Walter Bodmer, *The Public Understanding of Science.* London: Royal Society, 1985.

11. *Realising Our Potential: A Strategy for Science, Engineering and Technology.* London: Her Majesty's Stationery Office, 1993.

12. Arnold Wolfendale, *Report of the Committee to Review the Contribution of Scientists and Engineers to the Public Understanding of Science, Engineering and Technology.* London: Her Majesty's Stationery Office, 1996.

13. *Public Understanding of Science* (Bristol, U.K.: Institute of Physics Publishing) and *Science Communication* (Thousand Oaks, California: Sage).

14. E. D. Hirsch, *Cultural Literacy: What Every American Needs to Know.* Boston: Houghton Mifflin, 1987.

15. Richard P. Brennan, *Dictionary of Scientific Literacy.* Chichester: Wiley, 1992.

16. Robert M. Hazen and James Trefil, *Science Matters.* London: Cassell, 1993.

17. James Trefil, *1001 Things Everyone Should Know about Science.* London: Cassell, 1993.

18. Leon Lederman, Address to the International Conference on Public Understanding of Science and Technology, organized by the Chicago Academy of Sciences, Chicago, 1993.

19. Al Gore is quoted in *1996 Annual Report,* American Association for the Advancement of Science, Washington, D.C.

20. Geoffrey Thomas and John Durant, Why should we promote the public understanding of science? *Scientific Literary Papers 1* (1987), pp. 1–14.

21. U.S. President Bill Clinton is quoted on the front page of *The Times*, August 8, 1996.

22. Leon Lederman, *The God Particle.* London: Vintage, 1993); Steven Weinberg, *Dreams of a Final Theory.* London: Vintage, 1993.

23. Geofrrey A. Evans and John Durant, The relationship between knowledge and attitudes in the public understanding of science in Britain. *Public Understanding of Science 4* (1995), pp. 57–74.

24. Spencer Weart, *Nuclear Fear.* Cambridge, Massachusetts: Harvard University Press, 1988.

25. Jon Turney, *Frankenstein's Footprints—Biology and its Public in the Experimental Era.* Cambridge, Massachusetts: Harvard University Press, 1998.

26. Alan Shepard and Deke Slayton, *Moon Shot: The Inside Story of America's Race to the Moon.* London: Virgin, 1994.
27. James Gleick, *Genius: Richard Feynman and Modern Physics.* New York: Little, Brown, 1992.
28. Helen Lambert and Hilary Rose, Disembodied knowledge? Making sense of medical science. In: *Misunderstanding Science? The Public Reconstruction of Science and Technology,* edited by Alan Irwin and Brian Wynne. Cambridge: Cambridge University Press, 1996; pp. 65–83.
29. Susan B. Ruzek, *Feminist Alternatives to Medical Control.* New York: Praeger, 1978.
30. Brian Wynne, Public understanding of science. In: *Handbook of Science and Technology Studies,* edited by Sheila Jasanoff, Gerald E. Markle, James C. Petersen, and Trevor Pinch. Thousand Oaks, California: Sage, 1995; p. 380.
31. Frank Webster, *Theories of the Information Society.* London: Routledge, 1995.
32. Gerald Holton, *Science and Anti-Science.* Cambridge, Massachusetts: Harvard University Press, 1994.
33. C. P. Snow, *The Two Cultures.* Cambridge: Cambridge University Press, Canto Edition, 1993.
34. Matthew Arnold, Literature and science. In: *Philistinism in England and America,* edited by R. H. Super. Ann Arbor, Michigan: University of Michigan Press, 1974; pp. 53–73.
35. Bryan Appleyard, *Understanding the Present.* London: Picador, 1992.
36. I. Bernard Cohen, The education of the public in science. *Impact of Science on Society 3* (1952), pp. 78–81.
37. Leon E. Trachtman, The public understanding of science effort: a critique. *Science, Technology, & Human Values 1981* (Summer), pp. 10–12.

# CHAPTER 2

1. Hillier Krieghbaum, *Science and the Mass Media.* New York: New York University Press, 1967; pp. 18–19.
2. For essays on public science at this time, see "Science Lecturing in the Eighteenth Century," a special edition of the *British Journal for the History of Science* [vol. 28(1)], edited by Alan Q. Morton (Cambridge: Cambridge University Press, 1995).
3. Alan Q. Morton, *Science in the 18th Century: The King George III Collection.* London: Science Museum, 1993.

4. Morris Berman, *Social Change and Scientific Organisation: The Royal Institution, 1799–1844.* London: Heinemann Educational Books, 1978; pp. 100ff.

5. Robert V. Bruce, *The Launching of Modern American Science.* Ithaca, New York: Cornell University Press, 1987; p. 117.

6. Leslie Marchand, *The Athenaeum: A Mirror of Victorian Culture.* New York: Octagon, 1971.

7. Susan Holland and Steve Miller, Science in the early Athenaeum: A mirror of crystallisation. *Public Understanding of Science 6* (1997), pp. 111–130.

8. Ref. 5, p. 42 and Chapter 20.

9. Jack Morrell and Arnold Thackray, *Gentlemen of Science: Early Years of the British Association for the Advancement of Science.* Oxford: Clarendon Press, 1981; p. 22.

10. Richie Calder, Common understanding of science, *Impact of Science on Society 14* (1964), p. 181.

11. Bruce V. Lewenstein, Communiquer la science au public: l'emergence d'un genre americain, 1820–1939. In: *La science populaire dans la presse et l'édition, XIXè et XXè siècles.* Paris: CNRS Histoire, 1997; pp. 143–153.

12. Ref. 5, Chapter 19.

13. Robert M. Young, *Victorian Periodicals and the Fragmentation of a Common Context* in *Darwin's Metaphor.* Cambridge: Cambridge University Press, 1985.

14. John Tyndall, The Belfast Address, 1874. In: *Fragments of Science,* Vol. II. London: Longman, Green, 1892; pp. 135–201.

15. William Thomson, On the Age of the Sun's Heat. *Macmillan's Magazine,* March 1862, pp. 388–393.

16. M. J. Siefer, *Wizard: the Life and Times of Nikola Tesla, Biography of a Genius.* Secaucus, New Jersey: Birch Lane, 1996; p. 155.

17. Ref. 11.

18. Michael Shortland and Jane Gregory, *Communicating Science: A Handbook.* London: Longman, 1991; p. 14.

19. Ref. 11.

20. Ref. 11.

21. Ref. 18.

22. Ref. 11.

23. Ref. 18. p. 21.

24. Martin J. S. Rudwick, *The Great Devonian Controversy: The Shaping of Scientific Knowledge Among Gentleman Specialists.* Chicago: University of Chicago Press, 1985.

25. John C. Burnham, *How Superstition Won and Science Lost: Popularizing Science and Health in the United States.* New York: Rutgers University Press, 1987; p. 34.
26. Ref. 11.
27. Ref. 18.
28. See, for example, Peter Broks, *Media Science before the Great War.* London: Macmillan, 1996; in particular, pp. 111ff.
29. Geoffrey Martin, *Triumphs and Wonders of Modern Chemistry: A Popular Treatise on Modern Chemistry and Its Marvels, Written in Non-Technical Language for General Readers and Students.* London: Sampson Low, Marston, 1911.
30. Ibid., p. 64.
31. Alexander Findlay, preface to *Chemistry in the Service of Man,* 1st ed. (1916). London: Longman, Green, 1949; pp. ix–xi.
32. E. Frankland Armstrong (ed.), *Chemistry in the Twentieth Century: An Account of the Achievement and Present State of Knowledge in Chemical Science.* London: Ernest Benn, 1924; pp. 11–12.
33. Svante Arrhenius, *Chemistry and Modern Life.* London: Chapman & Hall, 1926; pp. 267–276.
34. Albert Fösling, *Albert Einstein: A Biography.* New York: Viking, 1997; pp. 445–446.
35. Ref. 11.
36. Ethel Mannin, *Young in the Twenties.* London: Hutchinson, 1970; p. 61.
37. Ref. 11.
38. Marcel LaFollette, *Making Science Our Own.* Chicago: University of Chicago Press, 1990; p. 53.
39. Ref. 11.
40. Edward Wyllis Scripps. *Encyclopaedia Britannica.* London: Encyclopaedia Britannica, 1929.
41. Ref. 18, p. 14.
42. Ref. 11.
43. Ref. 34, p. 458.
44. President's Research Committee on Social Trends 1933, *Recent Social Trends,* Vol. II. New York: McGraw-Hill, 1933; pp. 883–884.
45. Ibid., p. 903.
46. Ref. 36, p. 64.
47. Ref. 11.
48. Ian Ousby, *An Introduction to 50 American Novels.* London: Pan, 1979; p. 191.

49. Roslyn D. Haynes, *From Faust to Strangelove: Representations of the Scientist in Western Literature*. Baltimore: Johns Hopkins University Press, 1994; pp. 297–299; Sinclair Lewis, *Babbitt*. New York: Harcourt Brace Jovanovich and London: Cape, 1923; Sinclair Lewis, *Martin Arrowsmith*. New York: Buccaneer Books and London: Cape, 1925.

50. F. Scott Fitzgerald, *The Great Gatsby*. New York: Scribners, 1925.

51. John Dos Passos, *USA*. New York: Houghton Mifflin, 1930–36.

52. Ref. 36, p. 56. According to Roslyn D. Haynes, Huxley also presented science as destructive of social values and as responsible for the rise of nationalism. See Haynes, ref, 49. p. 205.

53. Brian Aldiss and David Wingrove, *Trillion Year Spree: The History of Science Fiction*. New York: Atheneum, 1986; p. 27.

54. Jane Banks and Jonathan David Tankel, Science as fiction: Technology in prime time television. *Critical Studies in Mass Communication 7* (1990), pp. 24–36.

55. Ref. 36, p. 27.

56. Ref. 18, p. 15.

57. Ref. 38, p. 45.

58. Ref. 36, p. 136.

59. Julian Huxley, *Memories*. London: George Allen and Unwin, 1970; p. 126.

60. Gary Werskey, *The Visible College*. London: Penguin, 1978; p. 85.

61. J. B. S. Haldane, *The Inequality of Man and Other Essays*. Harmondsworth: Penguin, 1938; p. 39.

62. Ibid., p. 125.

63. J. B. S. Haldane, How to write a popular scientific article. In: *On Being the Right Size*. Oxford: Oxford Paperback, 1985; pp. 154–160.

64. Lancelot Hogben, *Science for the Citizen*. London: Allen & Unwin, 1938; p. 10.

65. Ref. 59, p. 165.

66. This and subsequent quotes are from the Introduction to H. G. Wells, Julian Huxley, and G. P. Wells, *The New Science of Life*. (London: Cassell, 1931), p. 3.

67. Ref. 44, pp. 857–858.

68. Bruce V. Lewenstein, Industrial life insurance, public health campaigns, and public communication of science, 1908–1951. *Public Understanding of Science 1* (1992), pp. 347–365.

69. Ref. 18, p. 144.

70. Ref. 38, pp. 80–85.

71. Harlow Shapley, Samuel Rapport, and Helen Wright. (eds.), *A Treasury of Science.* New York: Harper & Brothers, 1943/1946; p. 7.

72. Cited in Daniel Kevles, *The Physicists: A History of a Scientific Community in Modern America.* Cambridge, Massachusetts: Harvard University Press, 1995; p. 377.

73. Robert C. Williams and Philip L. Cantelon. (eds.), *The American Atom: A Documentary History of Nuclear Policies from the Discovery of Fission to the Present, 1939–1984.* Philadelphia: University of Pennsylvania Press; p.161.

74. Anonymous, *Science in War.* Harmondsworth: Penguin, 1940.

75. Fred Hoyle, *Home Is Where the Wind Blows: Chapters from a Cosmologist's Life.* Mill Valley, California: University Science Books, 1994; p. 207.

76. Fred Hoyle, *A Decade of Decision.* Melbourne: William Heinemann, 1953; p. x.

77. J. C. Johnstone, An Astronomer Gets Down to Earth. *Daily Telegraph*, October 2, 1953.

78. Radical Surgery. *Economist,* October 10, 1953.

79. Ref. 18, p.15.

80. Quoted in Ref. 1, p. 4.

81. Martin Bauer, John Durant, Asdis Ragnarsdottir, and Annadis Rudolphsdottir, Science and Technology in the British Press, 1946–1990. London: The Science Museum, 1995; Vol. I, p. 8.

82. Ref. 10, p. 193.

83. Bruce V. Lewenstein, Public understanding of science in the United States after World War II, *Public Understanding of Science 1* (1992), pp. 55–57.

84. Spencer Weart, *Nuclear Fear: A History of Images.* Cambridge, Massachusetts: Harvard University Press, 1988; p. 100.

85. Stephen White, A newsman looks at physicists. *Physics Today 1* (1948), pp. 15–17, 33.

86. Ref. 81, pp. 12–16.

87. James Gleick, *Genius: Richard Feynman and Modern Physics.* New York: Little, Brown, 1992; pp. 340–341.

88. Hilary J. Deason, *A Guide to Science Reading.* New York: Signet, 1957–1964; p. vii.

89. Ref. 18, p. 15.

90. This and subsequent quotes are from contemporary editions of the *Radio Times,* the BBC's own listings magazine for radio and television.

91. For an analysis of these programs, see Marilee Long and Jocelyn Steinke, The thrill of everyday science: Images of science and scientists on children's ed-

ucational science programmes in the United States. *Public Understanding of Science 5* (1996), pp. 101–119.

92. For the story of *Discovering Women,* see Jocelyn Steinke, A portrait of a woman as a scientist: Breaking down barriers created by gender-role stereotypes. *Public Understanding of Science 6* (1997), pp. 409–428.

93. Ref. 18, p. 15.

94. This has been demonstrated for newspapers: see Ref. 81, pp. 16–18.

95. Christopher Dornan, Some problems of conceptualizing the issue of 'science and the media'. *Critical Studies in Mass Communication 7* (1990), pp. 48–49.

96. Rae Goodell, *The Visible Scientists.* Boston: Little, Brown, 1977.

97. See, for example, Dorothy Nelkin, *Selling Science: How the Press Covers Science and Technology* (revised edition). New York: W. H. Freeman, 1995; p. 114.

98. Ref. 38, p. 2.

99. C. P. Snow, *The Two Cultures.* Cambridge: Cambridge University Press, Canto Edition, 1993); with an introduction by Stefan Collini.

100. F. R. Leavis, The Significance of C. P. Snow. *The Spectator,* March 9, 1962, pp. 297–303.

101. Stephen Toulmin, *Cosmopolis: The Hidden Agenda of Modernity.* Chicago: University of Chicago Press, 1990.

102. Lionel Trilling, The Leavis–Snow controversy. In: *Beyond Culture: Essays on Literature and Learning.* New York: Longman, 1965; pp. 145–177.

103. T. H. Huxley, Science and culture. In: *Science and Education.* London: Macmillan, 1902; pp. 134–159.

104. Matthew Arnold, Literature and science. In: *Philistinism in England and America,* edited by R. H. Super. Ann Arbor, Michigan: University of Michigan Press, 1974; pp. 53–73.

105. Daniel Kevles, *The Physicists: A History of a Scientific Community in Modern America.* Cambridge, Massachusetts: Harvard University Press, 1955, p. 391.

106. Stefan Collini, introduction to C. P. Snow, *The Two Cultures.* Cambridge: Cambridge University Press, Canto Edition, 1993.

107. George Levine (ed.), One culture: Science and literature. In: *One Culture: Essays in Science and Literature.* Madison, Wisconsin: University of Wisconsin Press, 1987; pp. 3–28.

108. Fay Weldon, Thoughts We Dare Not Speak Aloud. *Daily Telegraph,* December 2, 1991, p. 14.

## CHAPTER 3

1. Stephen Hawking, *A Brief History of Time: From the Big Bang to Black Holes.* London: Bantam Press, 1988.
2. John Maddox, Defending science against anti-science. *Nature 368* (1994), p. 185.
3. Richard Dawkins, "The Richard Dimbleby Lecture," BBC1, November 12, 1996.
4. Gerald Holton, How to think about the 'anti-science' phenomenon. *Public Understanding of Science 1* (1992), pp. 103–128.
5. Paul. R. Gross and Norman Levitt, *Higher Superstition: The Academic Left and Its Quarrels with Science.* Baltimore: Johns Hopkins University Press, 1994; p. 2.
6. Ibid., p. 7.
7. Ref. 4, pp. 112–114.
8. Sergei Kapitza, Antiscience trends in the USSR. *Scientific American* 1991 (August), pp.18–24.
9. Harry Collins and Trevor Pinch, *The Golem: What Everyone Should Know about Science.* Cambridge: Cambridge University Press, Canto Edition, 1993; p. 2.
10. Ibid., pp. 57–78.
11. Robert K. Merton, Science and technology in a democratic order. *Journal of Legal and Political Sociology 1* 1942, pp. 115–126.
12. Bruno Latour, *Science in Action: How to Follow Scientists and Engineers through Society.* Cambridge, Massachusetts: Harvard University Press, 1987.
13. Ibid., pp. 13–14.
14. Ibid., p. 99.
15. David Bloor, *Knowledge and Social Imagery.* London: University of Chicago Press, 1976; 2nd ed., 1991.
16. Ibid., p. 7.
17. Ibid., p. 43.
18. Karl Popper, *The Logic of Scientific Discovery.* London: Hutchinson, 1959.
19. Thomas S. Kuhn, *The Structure of Scientific Revolutions.* Chicago: University of Chicago Press, 1962.
20. Ref., 15, pp. 55–83.
21. Ref. 15, p. 98.
22. See, for example, David Lyon, *Postmodernity* (Buckingham: Open University Press, 1994), for an accessible overview of postmodernism.

23. Jean-François Lyotard, *The Postmodern Condition: a Report on Knowledge,* translated by G. Bennington and B. Massumi. Manchester, U.K.: Manchester University Press, 1984.

24. Ibid., p. xxiv.

25. See, for example, Frederick Engels, *Anti-Dühring: Herr Eugen Dühring's Revolution in Science.* London: Lawrence & Wishart, 1969; and *The Dialectics of Nature.* Moscow: Progress Publishers, 1934.

26. J. D. Bernal, *Science in History.* London: Penguin Books, 1965.

27. Stanley Aronowitz, *Science as Power: Discourse and Ideology in Modern Society.* London: Macmillan Press, 1988.

28. Ibid., pp. 317–352.

29. Sandra Harding, *Whose Science? Whose Knowledge? Thinking from Women's Lives.* Milton Keynes, U.K.: The Open University Press, 1991; p. 3.

30. Ibid., p. 66.

31. Ibid., pp. 119ff.

32. Judy Wajcman, *Feminism Confronts Technology.* Cambridge: Polity Press, 1991; p. 162.

33. Ibid., pp. 81–109.

34. Aimee Sands, Never meant to survive: A black woman's journey—an interview with Evelynn Hammonds. In: *The "Racial" Economy of Science: Toward a Democratic Future,* edited by Sandra Harding. Bloomington: Indiana University Press, 1993; pp. 239–248.

35. Robert C. Johnson, Science, technology and black community development. In: *The "Racial" Economy of Science: Toward a Democratic Future,* edited by Sandra Harding. Bloomington: Indiana University Press, 1993; pp. 458–471.

36. Lloyd Timberlake, *Africa in Crisis: The Causes, the Cures of Environmental Bankruptcy.* London: Earthscan, 1991; p. 3.

37. Ibid., p. 72.

38. Rachel Carson, *Silent Spring.* London: Penguin Books, 1965; p. 22.

39. Ref. 1, p. 175.

40. Richard Dawkins, A Scientist's Case against God. *Independent,* April 20, 1992, p. 17.

41. John Habgood, Science and God: The Archbishop's Reply. *Independent,* May 4, 1992, p. 17.

42. John Polkinghorne, *Beyond Science: The Wider Human Context.* Cambridge: Cambridge University Press, 1996; pp. 75–94.

43. Bryan Appleyard, Science and the Spirit. *The Times Saturday Review,* April 25, 1992, pp. 12–13.

44. Bryan Appleyard, *Understanding the Present: Science and the Soul of Modern Man*. London: Picador, 1992; p. 9.

45. Ibid., pp. 1–16.

46. Matthew Arnold, Literature and science. In: *Philistinism in England and America*, edited by R. H. Super. Ann Arbor: The University of Michigan Press, 1974; pp. 53–73.

47. Mary Midgley, *Wisdom, Information and Wonder: What is Knowledge For?* London: Routledge, 1989; p. 45.

48. Carl Sagan, *The Demon-Haunted World: Science as a Candle in the Dark*. New York: Random House, 1996.

49. Peter W. Atkins, The Nobility of Knowledge. *Daily Telegraph,* May 11, 1992, p. 12.

50. Paul R. Gross, Norman Levitt, and Martin W. Lewis (eds.), *The Flight from Science and Reason*. Baltimore: Johns Hopkins University Press, 1997.

51. Alan Sokal, A physicist experiments with cultural studies. *Lingua Franca 1996* (May/June), pp. 62–64.

52. Barry Barnes, David Bloor, and John Henry, *Scientific Knowledge: A Sociological Analysis*. London: Athlone, 1996.

53. David Mermin, What's wrong with this sustaining myth? *Physics Today 49* (March 1996), pp. 11–13.; David Mermin, The golemization of relativity. *Physics Today 49* (April 1996), pp. 11–13.; Harry Collins, Trevor Pinch, and David Mermin, Sociologists, scientists continue debate about scientific process. *Physics Today 49* (July 1996), pp. 11–13.

54. Science Peace? Workshop, University of Southampton, July 28, 1997.

55. Michael Lynch, "Discovery, Construction, and the Science 'Peace Process'." Paper presented at the British Association annual meeting, September 11, 1997.

## CHAPTER 4

1. Anne Branscomb, Knowing how to know. *Science, Technology, & Human Values 1981* (Summer), p. 6.

2. Walter Bodmer, *The Public Understanding of Science*. London: Royal Society, 1986.

3. I. C. Jarvie, Media representations and philosophical representations of science. *Critical Studies in Mass Communications 7* (1990), p. 75.

4. Rae Goodell, *The Visible Scientists*. Boston: Little, Brown, 1977.

5. L. A. Lievrouw, Communication and the social representation of knowledge. *Critical Studies in Mass Communication 7* (1990), p. 4.

6. Bruce V. Lewenstein, personal communication.

7. Jeremy Dunning-Davies, Popular status and scientific influence: Another angle on 'the Hawking phenomenon'. *Public Understanding of Science 2* (1993), pp. 85–86. Dunning-Davies is an applied mathematician working in cosmology; he disagrees with Hawking's cosmology.

8. Ref. 5, p. 8.

9. Ref. 4.

10. Ref. 4., p.121.

11. T. H. Huxley, *Man's Place in Nature and Other Anthropological Essays.* London: Macmillan, 1897; p. ix.

12. Quoted in Hillier Krieghbaum, *Science and the Mass Media.* New York: New York University Press, 1967; p. 4.

13. Ref. 3, p. 72.

14. Harry M. Collins, The TEA set: Tacit knowledge and scientific networks. *Science Studies 4* (1974), p. 167.

15. Martin J. S. Rudwick, *The Great Devonian Controversy: the Shaping of Scientific Knowledge among Gentlemanly Specialists.* Chicago: University of Chicago Press, 1985; p. 430.

16. Ref. 5, pp. 2–3.

17. Susannah Hornig, Television's *NOVA* and the construction of scientific truth. *Critical Studies in Mass Communication 7* (1990), p. 12.

18. Richard Whitley, Knowledge producers and knowledge acquirers. In: *Expository Science: Forms and Functions of Popularization,* edited by Terry Shinn and Richard Whitley. Dordrecht: Reidel, 1985; pp. 3–10.

19. Ref. 5, p. 3.

20. Ref. 5, p. 9.

21. James E. Grunig, Communication of scientific information to nonscientists. In: *Progress in Communication Studies,* Vol. II, edited by B. Dervin and M. J. Voight. Norwood, New Jersey: Ablex, 1980; p. 185.

22. Stephen Hilgartner, The dominant view of popularization: Conceptual problems, political uses. *Social Studies of Science 20* (1990), pp. 519–539.

23. Marcel LaFollette, *Making Science our Own: Public Images of Science 1910–1955.* Chicago: University of Chicago Press, 1990; pp. 52–53.

24. Ref. 22, p. 531.

25. Bruce V. Lewenstein, From fax to facts: Communication in the cold fusion saga. *Social Studies of Science 25* (1995), p. 429.

26. Dudley Shapere, Scientific theories and their domains. In: *The Structure of Scientific Theories*, edited by Frederick Suppe. Urbana, Illinois: University of Illinois Press, 1977.

27. Daniel C. Dennett, *Consciousness Explained*. London: Penguin, 1993.

28. Baudouin Jurdant, Popularization of science as the autobiography of science. *Public Understanding of Science 2* (1993), pp. 365–373.

29. See Renaud Dulong and Werner Ackermann, Popularisation of science for Adults. *Social Science Information 11* (1972), p. 124.

30. Ref. 25, pp. 404–408.

31. Robert Logan, Popularization v. secularization: Media coverage of health. In: *Risky Business: Communicating Issues of Science, Risk and Public Policy*, edited by L. Wilkins and P. Patterson. Westport, Connecticut: Greenwood, 1992; p. 56.

32. Christopher Dornan, Some problems of conceptualizing the issue of 'science and the media'. *Critical Studies in Mass Communication*, 7 (1990), p. 51.

33. Ref. 22, p. 519.

34. J. Pradal, *The Literature of Science Popularization*. Strasbourg: Council of Europe, 1966.

35. Ref. 31, pp. 44–45.

36. Brian Wynne, Misunderstood misunderstandings: Social identities and the public uptake of science. *Public Understanding of Science 1* (1992), pp. 281–304.

37. Ref. 22, p. 528.

38. Ref. 25, pp. 408–424.

39. Niklas Luhmann, *Soziale Systeme. Grundiss einer allgemeinen Theorie.* Frankfurt: Suhrkamp, 1984.

40. Niklas Luhmann, What is communication? *Communication Theory 3* (1990), pp. 251–259.

41. See, for example, Hillier Krieghbaum, *Science and the Mass Media* (New York: New York University Press, 1967) and Geoffrey Thomas and John Durant, Why should we promote the public understanding of science?, in: *Scientific Literacy Papers* (Oxford: Department for Continuing Education, 1987), pp. 1–14.

42. Bruce V. Lewenstein, The meaning of 'public understanding of science' in the United States after World War II. *Public Understanding of Science 1* (1992), pp. 45–68.

43. John R. Durant, Geoffrey A. Evans, and Geoffrey P. Thomas, The public understanding of science. *Nature (London) 360* (1989), pp. 11–14.

44. Martin Bauer and Ingrid Schoon, Mapping variety in public understanding of science. *Public Understanding of Science 3* (1993), pp. 141–156.

45. John Durant, What is scientific literacy? In: *Science and Culture in Europe,* edited by John Durant and Jane Gregory. London: Science Museum, 1993; pp. 129–138.

46. Alan Chalmers, *What is This Thing called Science?* Milton Keynes, U.K.: Open University Press, 1978.

47. Henry H. Bauer, *Scientific Literacy and the Myth of the Scientific Method.* Chicago: University of Illinois Press, 1994.

48. Ibid., p. 151.

49. Harry Collins and Trevor Pinch, *The Golem: What Everyone Should Know about Science.* Cambridge: Cambridge University Press, Canto Edition, 1993.

50. Ref. 45, p. 136.

51. Morris H. Shamos, *The Myth of Scientific Literacy.* New Brunswick, New Jersey: Rutgers University Press, 1995.

52. Ibid., pp. 87–88, 215.

53. Ibid., pp. 192ff.

54. Robert Oppenheimer, *Science and the Common Understanding.* London: BBC Publications, 1953; p. 81.

55. Lewis Wolpert, *The Unnatural Nature of Science.* London: Faber, 1992.

56. Ibid., pp. 35ff.

57. This and subsequent quotes are from reviews by Michael Shortland, Jon Turney, and Harry Collins of Wolpert's *The Unnatural Nature of Science,* in *Public Understanding of Science 2* (1993), pp. 257ff.

58. Jürgen Habermas, *The Structural Transformation of the Public Sphere: An Inquiry into a Category of Bourgeois Society.* Cambridge: Polity Press, 1992.

59. Ibid., p. 27.

60. Denis McQuail, *Mass Communication Theory.* London: Sage, 1994.

61. Ref. 58, pp. 170–171.

62. Jean-Marc Levy-Leblond, About misunderstandings about misunderstandings. *Public Understanding of Science 1* (1992), pp. 17–21.

63. See Brian Wynne, The public understanding of science. In: *Handbook of Science and Technology Studies,* edited by Sheila Jasanoff, Gerald Markle, James C. Petersen, and Trevor Pinch. Thousand Oaks, California: Sage, 1995; p. 380.

64. Helen Lambert and Hilary Rose, Disembodied knowledge? Making sense of medical science. In: *Misunderstanding Science: The Public Reconstruction of Science and Technology,* edited by Alan Irwin and Brian Wynne. Cambridge: Cambridge University Press, 1996; pp. 65–83.

65. John Allen Paulos, *Innumeracy.* New York: Hill & Wang, 1988.

66. Ref. 36, pp. 271–294.

67. Robert Paine, 'Chernobyl' reaches Norway: The accident, science, and the threat to cultural knowledge. *Public Understanding of Science 1* (1992), pp. 261–270.

68. David Layton, Edgar Jenkins, Sally Macgill, and Angela Davey, *Inarticulate Science? Perspectives on the Public Understanding of Science and Some Implications for Science Education.* Leeds: Leeds Media Services, 1993.

69. Eugene Rabinowitch, Science popularization in the atomic age. *Impact of Science on Society 17*(2) (1967), p. 107.

70. See, for example, I. Bernard Cohen, Commentary: the fear and distrust of science in historical perspective. *Science, Technology, & Human Values 1981* (Summer), pp. 20–24; in 1997 an advertisement aimed at members of the British Association for the Advancement of Science for subscriptions to *Nature* read thus: "the misunderstanding of science is rife, not only amongst the so-called 'general public,' some of whom actually find science frightening . . . ."

71. See, for example, Ritchie Calder, Common understanding of science. *Impact of Science on Society 14* (1964), p. 179.

72. See, for example, Duncan Jackson, Bringing technology to the community: The Sellafield Visitors' Centre. In: *Museums and the Public Understanding of Science,* edited by John Durant. London: Science Museum, 1992; pp. 103–104.

73. See, for example, Geoffrey Evans and John Durant, The relationship between knowledge and attitude in the public understanding of science in Britain. *Public Understanding of Science 4* (1995), pp. 57–74; John Doble and Jean Johnson, *Science and the Public: A Report in Three Volumes.* New York: Public Agenda Foundation, 1990.

74. Friedhelm Neidhardt, The public as a communication system. *Public Understanding of Science 2* (1993), pp. 339–350.

75. Ref. 63, p. 380.

76. Ulrich Beck, *Risk Society: Towards a New Modernity.* London: Sage, 1992.

77. Anthony Giddens, *The Consequences of Modernity.* Cambridge: Polity, 1990.

## CHAPTER 5

1. Sharon Dunwoody, The challenge for scholars of popularized science: Explaining ourselves. *Public Understanding of Science 1* (1992), pp. 11–14.

2. For an introduction to media studies, see Denis McQuail, *Mass Communication Theory: An Introduction,* 2nd ed. London: Sage, 1994); and M. L. de Fleur and E. Dennis, *Understanding Mass Communication.* New York: Houghton Mifflin, 1994.

3. James E. Grunig, Communication of scientific information to nonscientists. In: *Progress in Communication Studies,* Vol. II, edited by B. Dervin and M. J. Voight. Norwood, New Jersey: Ablex, 1980; p. 169.

4. Christopher Dornan, Some problems in conceptualizing the issue of science and the media. *Critical Studies in Mass Communication 7* (1990), pp. 48–71.

5. Ibid., p. 49.

6. Dorothy Nelkin, *Selling Science: How the Press Covers Science and Technology* (revised edition) New York: W. H. Freeman, 1995.

7. Sharon Friedman, The journalist's world. In: *Scientists and Journalists: Reporting Science as News,* edited by Sharon Friedman, Sharon Dunwoody, and Carol Rogers (New York: Free Press, 1986), pp. 7–41; and Anders Hansen, Journalistic practices and science reporting in the British press. *Public Understanding of Science 3* (1994), pp. 111–134.

8. Sharon Dunwoody, The science writing inner club: A communication link between science and the lay public. In: *Scientists and Journalists: Reporting Science as News,* edited by Sharon Friedman, Sharon Dunwoody, and Carol Rogers. New York: Free Press, 1986; pp. 155–169.

9. See, for example, Anders Hansen, Journalistic practices and science reporting in the British press. *Public Understanding of Science 3* (1994), pp. 111–134.

10. There are many differing lists of news values, though they have much in common: this one is based on the work of Angela Stathopoulou and was itself developed from the work of Johan Galtung and Mari Ruge. See Johan Galtung and Mari Ruge, Structuring and selecting news. In: *The Manufacture of News: Deviance, Social Problems and the Mass Media,* edited by Stanley Cohen and Jock Young. London: Constable, 1973; pp. 2–73.

11. Jeanne Fahnestock, Accommodating science: The rhetorical life of scientific facts. In: *The Literature of Science—Perspectives on Popular Scientific Writing,* edited by Murdo William McRae. Athens, Georgia: University of Georgia Press, 1993; pp. 17–36.

12. Allan Bell, *The Language of News Media.* Oxford: Blackwell, 1991.

13. For a discussion of the potential and limitations of this type of research, see William Evans and Susanna Hornig-Priest, Science content and social context. *Public Understanding of Science 4* (1995), pp. 327–340.

14. Martin Bauer, John Durant, Asdis Ragnarsodottir, and Annadis Rudolphsdottir, *Science and Technology in the British Press, 1946–1990.* London: The Science Museum, 1995.

15. Marianne G. Pellechia, Trends in science coverage: A content analysis of three US newspapers. *Public Understanding of Science 6* (1997), pp. 49–68.

16. Ref. 13.

17. See, for example, Anders Hansen and Roger Dickinson, Science coverage in the British mass media: Media output and source input. *Communications 17*(3) 1992; pp. 365–377.

18. H. M. Collins, Certainty and the public understanding of science: Science on television. *Social Studies of Science 17* (1987), pp. 689–713; Susanna Hornig, Television's *NOVA* and the construction of scientific truth. *Critical Studies in Mass Communication 7* (1990), pp. 11–23.

19. Roger Silverstone, Narrative strategies in television science. In: *Impacts and Influences: Essays on Media Power in the Twentieth Century,* edited by James Curran, Anthony Smith, and Pauline Wingate. London: Methuen, 1987; pp. 291–330.

20. Richard Milton, *Forbidden Science: Suppressed Research That Could Change Our Lives.* London: Fourth Estate, 1994; pp. 135–143.

21. Conrad Smith, Reporters, news sources and scientific intervention: The New Madrid earthquake prediction. *Public Understanding of Science 5* (1996), pp. 205–216.

22. James W. Dearing, Newspaper coverage of maverick science: Creating controversy through balancing. *Public Understanding of Science 4* (1995), pp. 341–362.

23. David Gauntlett, *Moving Experiences: Understanding Television's Influences and Effects.* London: John Libbey, 1995.

24. Fiona Chew, Sushma Palmer, and Soohong Kim, Sources of information about health and nutrition: Can viewing one television programme make a difference? *Public Understanding of Science 4* (1995), pp. 17–30.

25. See the discussion of the issue in Edna Einsiedel, Framing science and technology in the Canadian press. *Public Understanding of Science 1* (1992), pp. 89–102.

26. Ref. 23, pp. 75–79.

27. See, for example, Michael Shortland, Networks of attitude and belief: Science and the adult student, in: *Scientific Literacy Papers,* edited by Michael Shortland. (Oxford: Department for External Studies, 1987), pp. 37–66; and Irene

Rahm and Paul Charbonneu, Probing Stereotypes through Students' Drawings of Scientists. *American Journal of Physics*, August (1997), pp. 774–777.

# CHAPTER 6

1. Thomas Henry Huxley, On a piece of chalk. In: *Discourses Biological and Geological.* London: Macmillan and Co., 1908; pp. 1–36.
2. Michael Faraday, *The Chemical History of a Candle.* London: Chatto and Windus, 1908; p. 14.
3. The lectures were repeated in 1860–61, and it is that version which has been published, as: Michael Faraday, *The Chemical History of a Candle.* London: Griffin, Bohn and Co., 1861.
4. Peter W. Atkins, *The Candle Revisited.* Oxford: Oxford University Press, 1994; p. 2.
5. Geoffrey Martin, *Triumphs and Wonders of Modern Chemistry: A Popular Treatise on Modern Chemistry and Its Marvels, Written in Non-Technical Language for General Readers and Students.* London: Sampson Low, Marston, 1911; p. 326.
6. E. Frankland Armstrong (ed.), introduction to *Chemistry in the Twentieth Century: An Account of the Achievement and the Present State of Knowledge in Chemical Science.* London: Ernest Benn, 1924; p. 11.
7. John Emsley, *The Consumer's Good Chemical Guide: A Jargon-free Guide to the Chemicals of Everyday Life.* Oxford: W. H. Freeman, 1994.
8. Jim Baggott, *Perfect Symmetry: The Accidental Discovery of Buckminster-fullerene.* Oxford: Oxford University Press, 1994. Hugh Aldersey-Williams, *The Most Beautiful Molecule: An Adventure in Chemistry.* London: Aurum Press, 1995.
9. A version of Steve Miller's analysis of this episode appeared as: Ten reasons why $C_{60}$ is such a good story. *Public Understanding of Science 6* (1997), pp. 207–212.
10. See, for example, the chapter entitled "What Makes a Good Science Story," in: *Scientists and Journalists,* edited by Sharon M. Friedman, Sharon Dunwoody, and Carol L. Rogers. (New York: The Free Press, 1986), pp. 103–116.
11. For an analysis of television science, see Roger Silverstone, Narrative strategies in television science. In: *Impacts and Influences: Essays on Media Power in the Twentieth Century,* edited by James Curran, Anthony Smith, and Pauline Wingate. New York: Methuen, 1987; pp. 291–330.

12. H. W. Kroto, J. R. Heath, S. C. O'Brien, R. F. Curl, and R. E. Smalley, $C_{60}$: Buckminsterfullerene. *Nature 318* (1985), pp. 162–163.

13. Tim Radford, Scientist's Just Dessert is Nobel Prize for Chemistry. *Guardian,* October 10, 1996, p. 7.

14. Ibid.

15. Alan J. Friedman and Carol C. Donley, *Einstein as Myth and Muse.* New York: Cambridge University Press, 1985.

16. See, for example, Ronald W. Clarke, *The Life and Times of Einstein* (New York: The World Publishing Company, 1985), Chapter 10; and Abraham Pais, Einstein and the press. *Physics World 1994* (August), pp. 30–36.

17. Harry Collins and Trevor Pinch, *The Golem: What Everyone Should Know about Science.* Cambridge: Cambridge University Press, Canto Edition, 1994; pp. 27ff.

18. Sir F. W. Dyson, A. S. Eddington, and C. Davidson, A determination of the deflection of light by the Sun's gravitational field from observations made at the total eclipse of May 29, 1919. *Philosophical Transactions of the Royal Society 220A* (1920), pp. 291–333.

19. Arthur S. Eddington, *Space, Time and Gravitation.* Cambridge: Cambridge University Press, 1920; pp. 110–122.

20. Abraham Pais, Einstein and the press. *Physics World 1994* (August), p. 31.

21. Ref. 15, p. 9.

22. Arthur I. Miller, *Insights of Genius: Imagery and Creativity in Science and Art.* New York: Springer-Verlag, 1996.

23. A version of Steve Miller's analysis of this episode appeared as: Wrinkles, ripples and fireballs: Cosmology on the front page. *Public Understanding of Science 3* (1994), pp. 445–454.

24. Marcus Chown, *Afterglow of Creation: From the Fireball to the Discovery of Cosmic Ripples.* London: Arrow Books, 1993; p. 5.

25. George Smoot and Keay Davidson, *Wrinkles in Time: The Imprint of Creation.* New York: Little, Brown, 1993; p. 273.

26. For more details, see Ref. 23.

27. Stephen Hawking, *A Brief History of Time: From the Big Bang to Black Holes.* London: Bantam, 1988; p. 175.

28. Marcus Wilkinson, If God is the Answer, What is the Question? *Financial Times Weekend Supplement,* April 18, 1992, p. 1.

29. Michael Rowan-Robinson, *Ripples in the Cosmos: A View behind the Scenes of the New Cosmology.* Oxford: W. H. Freeman, 1993; pp. 181–188.

30. John Polkinghorne, Scientists No Threat to God the Creator. *Sunday Times News Review,* April 26, 1992, p. 5.

31. Tom Wilkie, On a Butterfly Wing and a Prayer. *Independent,* May 12, 1992, p. 12.

32. Bernard Levin, Life, the Universe and Nothing. *The Times,* April 30, 1992, p. 12.

33. Tom Wilkie, Hunting for Holes in Hawking's Universe. *Independent on Sunday Magazine,* May 3, 1992, p. 52.

34. Sara Wheeler, All This and the Universe as Well. *Independent,* May 13, 1992, p. 14.

35. Robin McKie, Has Man Mastered the Universe? *Observer,* April 26, 1992, p. 4.

36. C. L. Bennet *et al.,* Preliminary separation of galactic and cosmic microwave emission for the COBE Differential Microwave Radiometer, *Astrophysical Journal Letters 396* (1992), pp. 7–12; E. L. Wright *et al.,* Interpretation of the cosmic microwave background anisotropy detected by the COBE Differential Microwave Radiometer, *Astrophysical Journal Letters 396* (1992), pp. 13–18; A. Kogut *et al.,* COBE Differential Microwave Radiometers: A preliminary systematic error analysis, *Astrophysical Journal 401* (1992), pp. 1–18.

37. Frederick Flam, COBE sows cosmological confusion. *Science 257* (1992), pp. 28–29.

38. Ref. 23.

39. John Durant, What is scientific literacy? In: *Science and Culture in Europe,* edited by John Durant and Jane Gregory. London: Science Museum, 1993; p. 129.

40. Anders Hansen, Greenpeace and press coverage of environmental issues. In: *The Mass Media and Environmental Issues,* edited by Anders Hansen. Leicester: Leicester University Press, 1993; p. 150.

41. Alison Abbott, Brent Spar: When science is not to blame. *Nature 380* (1996), p. 13.

42. James R. Bragg, Roger C. Prince, E. James Harner, and Ronald M. Atlas, Effectiveness of bioremediation for the Exxon Valdez oil spill. *Nature 368* (1994), pp. 413–417.

43. Richard P. J. Swannell and Ian M. Head, Bioremediation comes of age. *Nature 368* (1994), p. 396.

44. Henry Petroski, *The Pencil: A History.* New York: Alfred A. Knopf, 1989.

45. Friedhelm Neidhardt, The public as a communication system. *Public Understanding of Science 2* (1993), pp. 339–350.

## CHAPTER 7

1. Sheila Jasanoff, Civilization and madness: the great BSE scare of 1996. *Public Understanding of Science 6* (1997), pp. 221–232.
2. Susanna Hornig, Reading risk: Public response to print media accounts of technological risk. *Public Understanding of Science 2* (1993), p. 95.
3. The differences between scientists' and the public's conceptions of the same ideas has been investigated by social psychologists. Originating with Serge Moscovici's study of the public's ideas about psychoanalysis in France in the 1950s, social representation theory has become a useful tool in the study of the public understanding of science. For an introduction to this theory see Rob Farr, "Common sense, science and social representations," *Public Understanding of Science 2* (1993), pp. 189–204.
4. For an introduction to the psychology of risk, see Baruch Fischoff, Sarah Lichtenstein, Paul Slovic, Stephen L. Derby, and Ralph L. Keeney, *Acceptable Risk*. New York: Cambridge University Press, 1981; and Lee Wilkins and Philip Patterson (eds), *Risky Business: Communicating Issues of Science, Risk, and Public Policy*. New York: Greenwood Press, 1991.
5. JoAnn Valenti and Lee Wilkins, An ethical risk communication protocol for science and mass communications. *Public Understanding of Science 4* (1995), p. 179.
6. Ubrich Beck, *Risk Society: towards a New Modernity*. London: Sage, 1992; p. 58.
7. Ref. 2, p. 96.
8. The story of Alar told here is based on accounts by Kerry E. Rodgers and by Sharon Friedman and her colleagues. See Kerry E. Rodgers, Multiple meanings of Alar after the scare: Implications for closure. *Science, Technology, & Human Values 21* (1996), pp. 177–197; and Sharon M. Friedman, Kara Villamil, Robyn A. Suriano, and Brenda P. Egolf, Alar and apples: Newspapers, risk, and media responsibility. *Public Understanding of Science 5* (1996), pp. 1–20.
9. Tom Sanders and Peter Bazalgette, *The Food Revolution*. London: Bantam, 1992; p. 28.
10. This section draws heavily on the unpublished paper by John Durant and Jane Gregory, "How Do Agendas Affect Outcomes: The Case of BSE," which was presented at the American Association for the Advancement of Science Public Understanding of Science Symposium, "Communicating

Scientific Knowledge to the Public: Messages and Agendas," February 1992.

11. Tony Smith, Reassurances on bovine spongiform encephalopathy. *British Medical Journal 298* (1989), p. 625.

12. James Erlichman, Meat Eaters Swallow Food Poison Risk. *Guardian,* January 29, 1990, p. 3.

13. Paul Wilenius, Mad Cows' Disease Kills Cat. *Today,* May 11, 1990, p. 1.

14. Graham Brough, Scientific Proof: Mad Cow Link to Humans. *Today,* May 16, 1990, p. 1.

15. Geoffrey Cannon, Steve Connor, Helen Hague, and David Nicolson-Lord, Mad Cows and Englishmen. *Independent on Sunday,* May 20, 1990, p. 19.

16. Olga Craig and Wayne Francis, We Were Frightened Then Gummer Rang. *Today,* May 16, 1990, p. 5.

17. Stephen War, Loophole Allows Export of 'Mad Cow' Beef. *Independent,* June 9, 1990, p. 3.

18. 'Beef Safe'—Gummer. *Private Eye,* May 25, 1990.

19. David Nicholson-Lord, It's a Mad, Mad, Diet of MAFF-Speak. *Independent on Sunday,* September 30, 1990, p. 4.

20. John Pienaar and James Cusick, Farmers Attack Gummer on BSE. *Independent,* May 21, 1990, p. 1.

21. Michael Kenward, Mad as a Catter. *New Scientist,* May 19, 1990, p. 21.

22. See, for example, Tom Wilkie, Policy Decisions on BSE Made 'With Limited Evidence'. *Independent,* May 17, 1990, p. 5.

23. Editorial, Frankness needed. *Nature (London) 345* (1990), p. 648.

24. Richard Lacey, *Guardian,* June 13, 1990.

25. Peter Koenig, Mad cows no threat to a large vested interest. *Independent on Sunday,* May 20, 1990, p. 8.

26. Jane Doorly, letter to the *Independent on Sunday,* May 27, 1990, p. 30.

27. Ivor Crewe, Poll Shows Dearth of Trust in Ministry. *Independent,* May 17, 1990, p. 5.

28. Stephanie Pain, What Madness is This? *New Scientist,* June 9, 1990, p. 32.

29. Alex Renton, Doubts over BSE Safety Guarantee. *Independent,* May 23, 1990, p. 5.

30. Lisa DeProspo, An analysis of American newspaper articles regarding the link between BSE and Creutzfeldt–Jakob Disease. Unpublished paper, Department of Communication, Cornell University, April 1997.

31. Akiko Nishimura, Mad cow disease in Britain 1986–1996: The representation of a food problem in the press and among students. Unpublished paper, De-

partment of Social Psychology, London School of Economics and Political Science, July 1996.

32. Richard Rhodes, *Deadly Feasts: Tracking the Secrets of a Terrifying New Plague*. New York: Simon and Schuster, 1997.

33. Ref. 31, p. 21.

34. See, for example, the *Guardian*, March 21, 1996, and the *Independent*, March 21, 1996.

35. Ref. 1, p. 224.

36. Ibid. p. 223.

37. David Brown, Beef is Back to Pre-BSE Sales. *Daily Telegraph*, July 19, 1997, p. 1.

38. Richard Kimberlin, quoted in the *Independent*, March 24, 1990.

39. Richard Lacey, quoted in the *Independent*, March 24, 1990.

40. Ref. 1, p. 229.

41. See, for example, Nigel Calder, *The Comet Is Coming: The Feverish Legacy of Mr Halley* (London: Penguin, 1980), for a discussion of the history of cometary mythology.

42. Luis W. Alvarez, Walter Alvarez, Frank Asaro, and Helen Mitchell, Extraterrestrial cause for the Cretaceous–Tertiary extinction: Experimental results and theoretical interpretation. *Science 208* (1980), pp. 1095–1108.

43. See, for example, David M. Raup, *The Nemesis Affair: A Story of the Death of Dinosaurs and the Way of Science*. New York: Norton, 1986.

44. Clark R. Chapman and David Morrison, *Cosmic Catastrophes*. New York: Plenum Press, 1989; p. 277.

45. Elisabeth S. Clemens, The impact hypothesis and popular science: Conditions and consequences of interdisciplinary debate. In: *The Mass Extinction Debates: How Science Works in a Crisis*, edited by William Glen. Stanford: Stanford University Press, 1994; pp. 92–120.

46. Arthur C. Clarke, The Risk of Raining Stones. *Times Higher Educational Supplement*, August 22, 1997, p. 20.

47. Tom Gehrels, preface to *Hazards Due to Comets and Asteroids*. Tucson: University of Arizona Press, 1994.

48. Victor M. Clube, Comets: Hazards from space. In: *The Mass Extinction Debates: How Science Works in a Crisis*, edited by William Glen. Stanford: Stanford University Press, 1994; pp. 153–169.

49. Eugene H. Levy, Early impacts: The Earth emergent from its cosmic environment. In: *Hazards Due to Comets and Asteroids*, edited by Tom Gehrels. Tucson: University of Arizona Press, 1994; pp. 3–7.

50. David Morrison, *The Spaceguard Survey: Report of the NASA International Near-Earth Object Detection Workshop.* Pasadena, California: NASA/Jet Propulsion Laboratory, 1992.

51. Clark R. Chapman and David Morrison, Impacts on the Earth by asteroids and comets: Assessing the hazard. *Nature (London) 367* (1994), pp. 33–40.

52. Tim Radford, Serial Killers from Heaven. *The Guardian Online,* July 7, 1994, pp. 10–11.

53. Colin Macilwain and Maggie Verrall, Comet collision boosts controversy over global protection strategy. *Nature (London) 370* (1994), p. 165.

54. *Science* produced a special issue on March 3, 1995. *Science 267* (1995), p. 1245; pp. 1277–1323.

55. See, for example, John R. Spencer and Jacqueline Mitton (eds.), *The Great Comet Crash: The Collision of Comet Shoemaker-Levy 9 and Jupiter.* Cambridge: Cambridge University Press, 1995; Steve Miller, The impact of Comet Shoemaker-Levy 9 on Jupiter and on the world of astronomy, in: *The 1997 Yearbook of Astronomy,* edited by Patrick Moore. London: Macmillan, 1996; pp. 208–231; and Duncan Steel, *Rogue Asteroids and Doomsday Comets: The Search for the Million Megaton Menace That Threatens Life on Earth.* New York: Wiley, 1995.

56. David Morrison, review of *Rogue Comets and Doomsday Asteroids,* by Duncan Steel. *Physics Today 50* (February 1997), p. 65.

57. Ref. 4, Fischoff *et al.*, pp. 80–83.

58. Sharon Dunwoody and Hans Peter Peters, Mass media coverage of scientific and technological risks: A survey of research in the United States and Germany. *Public Understanding of Science 1* (1992), p. 221.

59. Ref. 9, p. 30.

60. Sarah Stewart-Brown and Andrew Farmer, Screening could seriously demage your health. *British Medical Journal 314* (1997), p. 533.

61. Alan Mazur, Technical risk in the mass media. *Risk: Health, Safety and Environment 189* (1994), p. 5.

62. Lynn J. Frewer, Maarten M. Raats, and Richard Shepherd, Modelling the media: The transmission of risk information in the British quality press. *IMA Journal of Mathematics Applied in Business & Industry 5* (1993/4), pp. 235–247.

63. Ref. 2, p. 98.

64. Friedhelm Neidhardt, The public as a communication system. *Public Understanding of Science 2* (1993), p. 347.

65. Ref. 6, p. 58

66. Ref. 64, p. 341.
67. Brian Wynne, Public understanding of science. In: *Handbook of Science and Technology Studies,* edited by Sheila Jasanoff, Gerald E. Markle, James C. Petersen, and Trevor Pinch. Thousand Oaks, California: Sage, 1995; p. 385.
68. Ref. 5, p. 184.
69. Ref. 5, pp. 177–194.
70. Philip J. Frankenfeld, Technological citizenship: A normative framework for risk studies. *Science, Technology, & Human Values 17* (1992), pp. 459–484.
71. Ref. 67, p. 377.
72. Ref. 2, p. 106.
73. Ref. 26.
74. J. S. Hargrave-Wright, letter to *New Scientist,* June 9, 1990, p. 76.
75. Ref. 64, p. 348.
76. See for example, Ref. 67, p. 380.
77. Ref. 1, p. 227.
78. Ibid., p. 228.
79. Ref. 2, p. 107.
80. Ref. 5, p. 187.

# CHAPTER 8

1. Robert V. Bruce, *The Launching of Modern American Science 1846–1876.* Ithaca, New York: Cornell University Press, 1987.
2. Roger Silverstone, The medium is the museum: On objects and logics in times and spaces. In: *Museums and the Public Understanding of Science,* edited by John Durant. London: Science Museum, 1992; p. 34.
3. Willem Hackmann, 'Wonders in one closet shut': The educational potential of history of science museums. In: *Museums and the Public Understanding of Science,* edited by John Durant. London: Science Museum, 1992; p. 65.
4. Georges Cuvier, *Rapports à l'Empereur sur le progrès des sciences, des lettres et des arts depuis 1789, II: Chimie et sciences de la nature.* Paris: Belin, 1989; pp. 295–296.
5. Daniel Kevles, *The Physicists: A History of a Scientific Community in Modern America.* Cambridge, Massachusetts: Harvard University Press, 1995; p. 63.
6. Stella Butler, *Science and Technology Museums.* Leicester: Leicester University Press, 1992; p. 8.
7. Ref. 3, p. 66.

8. Robert W. Rydell, *All the World's a Fair.* Chicago: University of Chicago, 1987; p. 4.

9. Ibid., p. 2.

10. William E. Cameron, Thomas W. Palmer, and Francis E. J. Willard, *The World's Fair, Being a Pictorial History of the Columbian Exposition.* Philadelphia: National Publishing Company, 1893.

11. Marc J. Siefer, *Wizard: The Life and Times of Nikola Tesla, Biography of a Genius.* Secaucus, New Jersey: Birch Lane, 1996; pp. 117–120.

12. Francis Bacon, *New Atlantis* (1667), cited in Richard Gregory, *Hands-On Science: An Introduction to the Bristol Exploratory.* London: Duckworth, 1986; pp. 26–27.

13. Ref. 3, p. 67.

14. I. B. Cohen, The education of the public in science. *Impact of Science on Society 3* (1952), p. 75.

15. Sheila Grinell, *A New Place for Learning Science: Starting and Running a Science Center.* Washington, D.C.: Association of Science-Technology Centers, 1992; pp. 6–7.

16. John G. Beetlestone, Colin H. Johnson, Melanie Quin, and Harry White, The science centre movement: Contexts, practice, next challenges. *Public Understanding of Science 7* (1998), pp. 5–26.

17. Graham Farmelo, Drama on the galleries. In: *Museums and the Public Understanding of Science,* edited by John Durant. London: Science Museum, 1992; pp. 45–49.

18. Ref. 6, p. 12.

19. Sylvia Chaplin and Francis Graham-Smith, Jodrell Bank Visitor Centre. In: *Museums and the Public Understanding of Science,* edited by John Durant. London: Science Museum, 1992; pp. 108–109.

20. Ref. 6, pp. 66–76.

21. Duncan Jackson, Bringing technology to the community: Sellafield Visitors' Centre. In: *Museums and the Public Understanding of Science,* edited by John Durant. London: Science Museum, 1992; pp. 103–107.

22. BNFL spokeswoman quoted in: Meg Carter, Adwatch: Public Relations Goes Nuclear. *Independent,* August 16, 1997, p. 18.

23. Melanie Quin, Clone, hybrid or mutant? The evolution of European science museums. In: *Science and Culture in Europe,* edited by John Durant and Jane Gregory. London: Science Museum, 1993; p. 196.

24. Ref. 6, p. 112.

25. Alfred de Grazia, *The Velikovsky Affair: Scientism versus Science.* London: Sphere, 1978; p. 30.

26. Joel Bloom, Science and technology museums face the future. In: *Museums and the Public Understanding of Science,* edited by John Durant. London: Science Museum, 1992; p. 19.

27. Sharon Macdonald, Authorising science: Public understanding of science in museums. In: *Misunderstanding Science? The Public Reconstruction of Science and Technology,* edited by Alan Irwin and Brian Wynne. Cambridge: Cambridge University Press, 1996; p. 152.

28. Ibid.

29. Ref. 26, pp. 16–17.

30. Ibid., p. 18.

31. Ref. 6, p. 113.

32. Ref. 6, p. xi.

33. Creative Research, The Geological Museum—a Market Research Study, report to the management of the Natural History Museum, London (1994); The Susie Fisher Group, Bringing Gemstones and Minerals Alive, report to the management of the Natural History Museum (1994).

34. The information for this example came from our own visits and discussions with staff at the Natural History Museum, especially Dr. Robert Bloomfield, who was in overall charge of the transformation of the Earth Galleries.

35. Ref. 1, p. 339.

36. Series Preface to Stella Butler, *Science and Technology Museums.* Leicester: Leicester University Press, 1992; p. vii.

37. Roger Miles, Museums and the communication of science. In: *Communicating Science to the Public,* edited by The Ciba Foundation. Chichester: Wiley, 1987; p. 116.

38. Ref. 26, p. 18.

39. Ref. 16, p. 15.

40. Ref. 16, p. 8

41. Ref. 37, p. 117.

42. For a survey of visitor studies research, see Roger Miles and Alan Tout, Exhibitions and the public understanding of science. In: *Museums and the Public Understanding of Science,* edited by John Durant. London: Science Museum, 1992; pp. 27–33.

43. Gaynor Kavanagh, Dreams and nightmares: Science museum provision in Britain. In: *Museums and the Public Understanding of Science,* edited by John Durant. London: Science Museum, 1992; p. 65.

44. Ref. 16, p. 18.

45. Ref. 16, pp. 18–19.

46. Ref. 42, p. 27.

47. See, for example, P. Anderson, with B. Cook Roe, *Museum Impact and Evaluation Study*. (Chicago: Museum of Science and Industry, 1993) and H. Salmi, Science Centre Education: Motivation and Learning in Informal Education, Department of Teacher Education, University of Helsinki, Research Report 119 (1993), pp. 1–202.

48. Ref. 42, p. 28.

49. Alan Friedman, quoted in discussion in Ref. 37, p. 123.

50. Burton Richter of the American Physical Society is quoted in Bruce V. Lewenstein, Shooting the messenger: Understanding attacks on *Science in American Life*. Paper presented to the 4th International Conference on Public Communication of Science and Technology, Melbourne, Australia, November 11, 1996.

51. Alan Friedman, Exhibits and expectations. *Public Understanding of Science 4* (1995), p. 306.

52. Robert Park, Is science the god that failed? *Science Communication 16*(2) (1994), p. 207.

53. Ibid., p. 208.

54. Ref. 51, p. 308.

55. Ref. 52, p. 209

56. Cited in Bruce V. Lewenstein, Shooting the messenger: Understanding attacks on *Science in American Life*. Paper presented to the 4th International Conference on Public Communication of Science and Technology, Melbourne, Australia, November 11, 1996.

57. Ref. 51, p. 311.

58. Ref. 52, p. 210.

59. Ref. 56, p. 4.

60. Ibid., p. 16.

61. Ref. 51, p. 312.

# CHAPTER 9

1. Previous attempts at this sort of overview provided a starting point for this chapter: see John Durant and Jane Gregory (eds.), *Science and Culture in Europe* (London: Science Museum, 1993) and Bernard Schiele (ed.), *When Science Becomes Culture: World Survey of Scientific Culture* Boucherville, Quebec: University of Ottawa Press, 1994).

2. American Association for the Advancement of Science, *Project 2061: Science for All Americans.* New York: Oxford University Press, 1994.

3. National Science Foundation, *NSF in a Changing World: The National Science Foundation's Strategic Plan, 1995.* Washington, D.C.: National Science Foundation, 1995.

4. Ibid., p. 18.

5. Membership Report, Association of Science-Technology Centers, Washington, D.C., 1992. John G. Beetlestone, Colin H. Johnson, Melanie Quin, and Harry White, The science centre movement: contexts, practice, next challenges. *Public Understanding of Science 6* (1997), pp. 5–26.

6. National Science Board, *Science and Engineering Indicators—1996.* Washington, D.C.: National Science Board, 1997.

7. Gillian Pearson, Susan M. Pringle, and Jeffrey N. Thomas, Scientists and the public understanding of science, *Public Understanding of Science 6* (1997), pp. 270–290.

8. Grahamstown Foundation, *From Space to Cyberspace: SciFest 1997 Report.* Grahamstown, South Africa: Grahamstown Foundation, 1997.

9. Bruce V. Lewenstein, A survey of activities in public communication of science and technology in the United States. In: *When Science Becomes Culture: World Survey of Scientific Culture,* edited by Bernard Schiele. Boucherville, Quebec: University of Ottawa Press, 1994; pp.137–140.

10. European Trends: A Comparative Review. In: *When Science Becomes Culture: World Survey of Scientific Culture,* edited by Bernard Schiele. Boucherville, Quebec: University of Ottawa Press, 1994; pp. 351–352; and Joan Solomon, School science and the future of scientific culture. *Public Understanding of Science 5* (1996), pp. 157–165.

11. Edgar W. Jenkins (ed.), *Innovations in Science and Technology Education,* Vol. VI. Paris: UNESCO, 1997.

12. A. O. Kuku, Science and technological literacy. In: *Innovations in Science and Technology Education,* Vol. VI, edited by Edgar W. Jenkins. Paris: UNESCO, 1997; p. 141.

13. John Burns, Girls, women and scientific and technological literacy. In: *Innovations in Science and Technology Education,* Vol. VI, edited by Edgar W. Jenkins. Paris: UNESCO, 1997; pp. 125–139.

14. Bernard Schiele (ed.), *When Science Becomes Culture: World Survey of Scientific Culture* Boucherville, Quebec: University of Ottawa Press, 1994.

15. Anders Hansen (ed.), *The Mass Media and Environmental Issues.* Leicester: Leicester University Press, 1993). See particularly chapters by Jon Cracknell,

Issue arenas, pressure groups and environmental agendas, pp. 3–21, and Anders Hansen, Greenpeace and press coverage of environmental issues, pp. 150–178.

16. Alison Goddard, Fighting science with science. *Physics World* 7 (1997), pp. 57–58.

17. Susan B. Ruzek, *Feminist Alternatives to Medical Control.* New York: Praeger, 1978.

18. Steven Epstein, *Impure Science: AIDS Activism and the Politics of Knowledge.* Berkeley: University of California Press, 1996; pp. 336–336.

19. See, for example, John Kluger, Calling All Amateurs. *Time Magazine,* August 11, 1997, p. 44.

20. John Hartman, The popularisation of science through citizen volunteers. *Public Understanding of Science* 6 (1997), pp. 69–86.

21. John Gilles, *Media Fellowship Report.* London: British Association for the Advancement of Science, 1992.

22. Alan Derrington, quoted in *Going Public: An Introduction to Communicating Science, Engineering and Technology.* London: Department of Trade and Industry, 1997; p. 8.

23. Byron H. Waksman, Bernhard C. Adelmann-Grill, and Georg W. Kreutzberg, EICOS: A European initiative for communicators of science," *Public Understanding of Science* 2 (1993), pp. 245–256.

24. Alan H. McGowan, The significance of the media resource service—now services. In: *The Role of the Media in Science Communication.* London: The CIBA Foundation, 1993; pp. 51–61.

25. The CIBA Foundation, *Report and Handbook.* London: CIBA Foundation, 1985.

26. Klaus Meier, The world's favourite experts. *CIBA Foundation MRS Newsletter 1977* (Spring), pp. 1–3.

27. Richard Wiseman, 'MegaLab UK': Participatory science and the mass media. *Public Understanding of Science* 5 (1996), pp. 167–169.

28. Pierre Fayard, Making science go round the public. In: *When Science Becomes Culture: World Survey of Scientific Culture,* edited by Bernard Schiele. Boucherville, Quebec: University of Ottawa Press, 1994; pp. 357–380.

29. Alan Irwin, *Citizen Science: A Study of People, Expertise and Sustainable Development.* London: Routledge, 1995.

30. Marcel P. R. van den Broeke, Science information centres and science shops: Information on demand through new media. In: *The Role of the Media in Science Communication.* London: The CIBA Foundation, 1993; pp.75–86.

31. See, for example, Judith Petts, The public–expert interface in local waste management decisions: Expertise, credibility and process. *Public Understanding of Science 6* (1997), pp. 359–381.

32. See, for example, Simon Joss and John Durant (eds.), *Public Participation in Science: The Role of Consensus Conferences in Europe.* London: The Science Museum, 1995.

33. J. Grundahl, The Danish consensus conference model. In: *Public Participation in Science: The Role of Consensus Conferences in Europe,* edited by Simon Joss and John Durant. London: The Science Museum, 1995; pp. 31–40. L. Kluver, Consensus conferences at the Danish Board of Technology. In: *Public Participation in Science: The Role of Consensus Conferences in Europe,* edited by Simon Joss and John Durant. London: The Science Museum, 1995; pp. 41–49.

34. The lay panel of the first U.K. consensus conference came down in favor of genetic engineering of plants so long as particular safeguards were in place. See Simon Joss and John Durant, The UK National Consensus Conference on Plant Biotechnology. *Public Understanding of Science 4* (1995), pp. 195–204.

35. For a fuller discussion of mechanisms for involving the public and the thinking behind them, see Simon Joss, Participation in parliamentary technology assessment: From theory to practice. In: *Parliaments and Technology: The Development of Technology Assessment in Europe,* edited by N. J. Vig and H. Paschen. New York: New York State University Press, 1998.

36. For more information about Loka Alerts, contact the Loka Institute in Amherst, Massachusetts, at loka@amherst.edu, or see http://www.amherst.edu/~loka. This extract is from "Science Inc. vs. Science for Everyone," the trial "Loka Alert," dated July 24, 1997, by Dick Sclove.

37. Jon Turney, *To Know Science Is to Love It? Observations from Public Understanding of Science Research.* London: COPUS, 1997.

38. Jon D. Miller, The relationship between biomedical understanding and public policy. Paper for *Education for Scientific Literacy Conference.* London: Science Museum, 1994.

39. John Doble and Jean Johnson, *Science and the Public: A Report in Three Volumes.* New York: The Public Agenda Foundation, 1990.

40. Sharon Dunwoody, Elizabeth Crane, and Bonnie Brown, *Directory of Science Communication Courses and Programs in the United States.* Madison, Wisconsin: Center for Environmental Communications and Education Studies, University of Wisconsin, 1996.

41. Jane Randall Crockett, Training and development for informal science learning. *Public Understanding of Science 6* (1997), pp. 87–102.

42. Jon Turney, Teaching science communication: Courses, curricula, theory and practice. *Public Understanding of Science 3* (1994), pp. 435–444.


## CHAPTER 10

1. Harry Collins, review of *The Unnatural Nature of Science*, by Lewis Wolpert. *Public Understanding of Science 2* (1993), pp. 261–264.

2. Geoffrey Thomas and John Durant, Why should we promote the public understanding of science? In: *Scientific Literacy Papers*. Oxford: Department for External Studies, University of Oxford, 1987; pp. 1–14.

3. H. G. Wells, quoted in Julian Huxley, *Memories*. London: George Allen and Unwin, 1970; p. 165.

4. See, for example, David Lindsay, *A Guide to Scientific Writing: Manual for Students and Research Workers* (Melbourne: Longman Cheshire, 1984) and Maeve O'Connor, *Writing Successfully in Science* (London: Harper Collins, 1991), for manuals for writing scientific papers and theses; and, for example, Stephen White, Peter Evans, Chris Mihill, and Maryon Tysoe, *Hitting the Headlines: A Practical Guide to the Media* (Leicester: The British Psychological Society, 1993) and Michael Shortland and Jane Gregory, *Communicating Science: A Handbook* (New York: Wiley, 1991), for guides to communicating through the mass media and directly with the public.

5. See, for example, Tim Albert, *Medical Journalism: The Writer's Guide.* (Oxford: Radcliffe Medical Press, 1992), Deborah Blum and Mary Knudson (eds.), *A Field Guide for Science Writers: The Official Guide of the National Association of Science Writers* (New York: Oxford University Press, 1997), and Sharon M. Friedman, Sharon Dunwoody, and Carol L. Rogers (eds.), *Scientists and Journalists: Reporting Science as News* (New York: The Free Press, 1986).

6. See, for example, Victor Cohn, Coping with statistics. In: *A Field Guide for Science Writers: The Official Guide of the National Association of Science Writers*, edited by Deborah Blum and Mary Knudson. New York: Oxford University Press, 1997; p. 103.

# Index